12/93 Lauren - To ♡ W9-BRP-425
little 'math whiz'

Love - mom

xox

Also by Arthur Benjamin and Michael Shermer:

Teach Your Child Math

Also by Michael Shermer:

Psychling: On Mental and Physical Fitness Sport Cycling

The Ultra-Marathon Cycling Manual (with John Marino and Lon Haldeman)

Cycling: Endurance and Speed

The Woman Cyclist (with Elaine Mariolle)

The RAAM Book (with John Marino and Lon Haldeman)

Teach Your Child Science

Meeting the Challenge of Arthritis (with George Yates)

MATHEMAGICS

HOW TO LOOK LIKE A GENIUS WITHOUT REALLY TRYING

· · · ·

Arthur Benjamin, Ph.D., *and*
Michael Brant Shermer, Ph.D.

LOWELL HOUSE
Los Angeles
CONTEMPORARY BOOKS
Chicago

Library of Congress Cataloging-in-Publication Data

Benjamin, Arthur.
   Mathemagics : how to look like a genius without
really trying / Arthur Benjamin and Michael
Shermer.
     p.   cm.
   Includes bibliographical references and index.
   ISBN 0-929923-54-5
   1. Mathematical recreations.   I. Shermer,
Michael.   II. Title.
QA95.B444   1993
793.7'4—dc20   92-16834
                              CIP

Requests for such permissions should be
addressed to:

     Lowell House
     2029 Century Park East, Suite 3290
     Los Angeles, CA 90067

Publisher: JACK ARTENSTEIN
Vice-President/Editor-in-Chief: JANICE GALLAGHER
Director of Publishing Services: MARY D. AARONS
Illustrations: GREG LEVIN
Text Design: MIKE YAZZOLINO

Manufactured in the United States of America

10 9 8 7 6 5 4 3 2 1

ACKNOWLEDGMENTS

Special thanks go to Betsy Amster, who refined our efforts into a clean and concise text; Victoria Pasternack, for reminding us of who our audience is; Janice Gallagher, for seeing the project through to the end; Greg Levin, for illustrating the book in a manner that conveys the fun we have doing mathemagics; and Mary Aarons and Peter Hoffman, for producing a book of high quality that we can be proud of. Arthur Benjamin especially wants to acknowledge those who inspired him to become both a mathematician and a mathemagician—mathematicians Alan J. Goldman and Edward R. Scheinerman, magicians James Randi and Paul Gertner, mathemagician Martin Gardner; and Deena Dizengoff, for overall inspiration.

I dedicate this book to my parents, Larry and
Lenore Benjamin—
for everything.

ARTHUR BENJAMIN

My dedication is to my wife, Kim, for being my most
trusted confidante and personal counselor.

MICHAEL SHERMER

Contents

Foreword

Mathematics is a wonderful, elegant, and exceedingly useful language. It has its own vocabulary and syntax, its own verbs, nouns, and modifiers, and its own dialects and patois. It is used brilliantly by some, poorly by others. Some of us fear to pursue its more esoteric uses, while a few of us wield it like a sword to attack and conquer income tax forms or masses of data that resist the less courageous. This book does not guarantee to turn you into a Leibniz, or put you on stage as a Professor Algebra, but it will, we hope, bring you a new, exciting, and even entertaining view of what can be done with that wonderful invention—numbers.

We all think we know enough about arithmetic to get by, and we certainly feel no guilt about resorting to the handy pocket calculator that has become so much a part of our lives. But, just as photography may blind us to the beauty of a Vermeer painting, or an electronic keyboard may make us forget the magnificence of a Horowitz sonata, too much reliance on technology can deny us the pleasures that you will find in these pages.

I remember the delight I experienced as a child when I was shown that I could multiply by 25 merely by adding two zeros to my number and dividing by four. Casting out nines to check multiplication came next, and then when I found out about cross-multiplying I was hooked and became, for a short while, a generally unbearable math nut. Immunizations against such afflictions are not available. You have to recover all by yourself. Beware!

This is a fun book. You wouldn't have it in your hands right now if you didn't have some interest either in improving your math skills or in satisfying a curiosity about this fascinating subject. As with all such instruction books, you may only retain and use a certain percentage of the varied tricks and methods described here, but that alone will make it worth the investment of your time.

I know both the authors rather well. Art Benjamin is not only one of those whiz kids we used to groan over in school, but has even been known to tread the boards at the Hollywood Magic Castle, performing demonstrations of his skill, and on one occasion he traveled to Tokyo, Japan, to pit his math skills against a lady savant on live television. Michael Shermer, with his specialized knowledge of science, has an excellent overview of practical applications of math as it is used in the real world.

If this is your first exposure to this kind of good math stuff, I envy you. You'll discover, as you come upon each delicious new way to attack numbers, that you missed something in school. Mathematics, particularly arithmetic, is a powerful and dependable tool for day-to-day use which enables us to handle our complicated lives with more assurance and accuracy. Let Art and Michael show you how to round a few of the corners and cut through some of the traffic. Remember these words of Dr. Samuel Johnson, an eminently practical soul in all respects: "Arithmetical inquires give entertainment in solitude by the practice, and reputation in public by the effect."

Above all, enjoy the book. Let it entertain you, and have fun with it. That, with the occasional good deed, a slice of pizza (no anchovies!), and a selection of good friends, is about all you can ask of life. Well, almost all. Maybe a Ferrari . . .

James Randi
Plantation, Florida
April 1993

Preface

Dr. Arthur Benjamin, mathematics professor from Harvey Mudd College in Claremont, California, takes the stage to a round of applause at the Magic Castle, a celebrated magic club in Hollywood, where he is about to perform "mathemagics," or what he calls "the art of rapid mental calculation." Art appears nothing like a mathematics professor from a prestigious university. Astonishingly quick-witted, he looks at home with the rest of the young magicians playing at the Castle—which he is. Art has the unique distinction of being both a brilliant mathematician published in scholarly journals and a "lightning calculator," a recognized mathemagician performing at professional venues.

What makes Art so special is that he can play in front of any group, including professional mathematicians and magicians, because he can do something that almost no one else can. Art Benjamin can add, subtract, multiply, and divide numbers in his head faster than most people can with a calculator. He can square 2-digit, 3-digit, and 4-digit numbers, as well as find square roots and cube roots, without writing anything down on paper. And he can teach you how to perform your own mathematical magic.

Traditionally, magicians refuse to disclose how they perform their tricks. If they did, everyone would know how they are done and the mystery and fascination of magic would be lost. But Art wants to get people excited about math. And he knows that one of the best ways to do it is to let you and other readers in on the secrets of mathemagics. With these skills, almost anyone can do what Art Benjamin does every time he gets on stage to perform his magic.

This particular night at the Magic Castle, Art begins by asking if anyone in the audience has a calculator. A group of engineers raise their hands and join Art on stage. Offering to test their calculators to make sure they work, Art asks a member of the audience to call out a 2-digit number. "Fifty-seven," shouts one. Art points to another who yells out "23."

Directing his attention to those on stage, Art tells them: "Multiply 57 by 23 on the calculator and make sure you get 1311 or the calculators aren't working correctly." Art waits patiently while the volunteers finish inputting the numbers. As each participant indicates his calculator reads 1311, the audience lets out a collective gasp. The amazing Art has beaten the calculators at their own game!

Art next informs the audience that he will square four 2-digit numbers faster than his stage button-pushers can square them on their calculators. The audience members call out the numbers 24, 38, 67, 97, while Art jots the answers down on the blackboard on the stage as quickly as they can call them out. In large, bold letters for everyone to see, the sequence reads: 576, 1444, 4489, 9409. Art turns to his aides who are just now finishing their calculations and asks them to call out their answers. Their response triggers gasps and then applause from the audience: "576, 1444, 4489, 9409." The girl next to me sits with her mouth open in amazement.

Anticipating the audience's next response, Art explains: "I know what you're thinking: 'Oh, he just memorized all those numbers.' Well, it's true there are only ninety 2-digit numbers and I have been doing these problems for a good 10 years, so inadvertently I may have memorized a lot of the answers. But let me go a step farther and do some 3-digit calculations, almost none of which I've memorized. I won't even write them down because it will slow me up. I'll just square them in my head and call out the answer. If someone will give me a 3-digit number I will try to race you to the answer."

"Five hundred and seventy-two," a gentleman calls out. Art's reply comes less than a second later: "572 will give you 327,184." He immediately points to another member of the audience, who yells, "389," followed by Art's unblinking response: "389 squared will give you 151,321." Someone else blurts out "262." "That'll give you 68,644." Sensing he delayed just an instant on that last one, he promises to make up for it on the next number. The challenge comes—990. With no pause, Art gives the answer: "990 squared is 980,100." Several more 3-digit numbers are given and Art responds perfectly. Members of the audience shake their heads in disbelief.

With the audience in the palm of his hand, Art now declares that he will attempt a 4-digit number. A woman calls out, "1,036," and Art instantly responds "that's 1,073,296." The audience laughs and Art explains, "No, no, that's much too easy a number. I'm not supposed to beat the calculators on these. Let's try another one." A man offers a challenging 2843. Pausing briefly between digits, Art responds: "Let's see, 2,843 gives you 8 million . . . 82 thousand . . . 649." He is right, of course, and the audience roars their approval, as loudly as they did for the previous magician who sawed a woman in half and made a dog disappear.

Art explains that these calculations are the easy ones because when you square a number you only have to deal with one number, not two different numbers. To demonstrate that two different 4-digit numbers are harder to multiply mentally than squaring a single 4-digit number, Art asks that someone give him two 4-digit numbers from those already written down on the blackboard. The audience chooses the two largest numbers: 4489 and 9409. Clearly Art has his work cut out for him. "Uh . . . 42 million . . . 237 thousand and 1?" The re-sounding applause erases his vocal question mark. The mathemagician has escaped again.

It is the same everywhere Art Benjamin goes, whether it is a high school auditorium, a college classroom, a faculty lounge, the Magic Castle, or a television studio. Professor Benjamin has performed his special brand of magic live all over the country and on numerous television talk shows. He has been the subject of investigation by a cognitive psychologist at Carnegie Mellon University and is featured in a scholarly book by Steven Smith called *The Great Mental Calculators: The Psychology, Methods, and Lives of Calculating Prodigies.* Art was born in Cleveland on March 19, 1961 (which he calculates was a Sunday, a skill he will teach you in Chapter 9). A hyperactive child, Art drove his teachers mad with his classroom antics, which included correcting the mathematical mistakes they occasionally made. Through-out this book when teaching you his mathemagical secrets, Art recalls when and where he learned these skills, so I will save the fascinating stories for him to tell you.

Art Benjamin is an extraordinary individual with an extraordi-nary program to teach you rapid mental calculation. I offer these claims without hesitation, and only ask that you remember this does not come from a couple of guys promising miracles if you will only call our 800 hotline. Art and I are both credentialed in the most conservative of academic professions—Art in mathematics and I, myself, in the history of science—and we would never risk career embarrassment (or worse) by making such powerful claims if they were not true. To put it simply, this stuff works, and virtually everyone can do it because mathemagics is a learned skill. So you can look forward to improving your math skills, impressing your friends, enhancing your memory and, most of all, having fun!

Michael Shermer
Altadena, California, 1993

The Art of Rapid Mental Calculation

For as long as I can remember, mental arithmetic has kept me happily occupied. Getting the right answer to a difficult mathematical problem gives me the same thrill that others get from playing sports and games, reading books, watching movies, or solving crossword puzzles. Magic is my other passion, in large part because I've always been a ham. I've performed as a magician since I was a kid. Somewhere along the way, I learned to combine math and magic, and I have made it part of my show.

If you are like me, you will find that turning math into magic makes math a lot more fun. Once you know how, you can solve large arithmetic problems quickly and accurately in a matter of seconds. You can estimate the answers to problems of any size. And you can easily triple your ability to memorize numbers. In this book I reveal *all* of my mathemagical secrets. I've held nothing back.

Specifically, with *Mathemagics* by your side, you can learn to impress your friends, colleagues, and classmates with your amazing mathemagical skills. Once you have mastered the techniques within, you will be able to add, subtract, multiply, and divide faster than someone next to you with a calculator. This is the sort of skill that not only makes waves at parties or at gatherings of friends but also at the office, in school, and at banks, restaurants, or anywhere else where numbers matter. If you are afraid of math, this book can help you overcome your fears. If you are already pretty handy with numbers, *Mathemagics* will boost your abilities far beyond what you ever imagined.

The key to becoming a mathemagician is to ease your way into it. That is why the first four chapters of *Mathemagics* deal with basic mental computations. In Chapter 1 you will learn to add and subtract 2-digit, 3-digit, and even some 4-digit numbers in your head. Chapters 2 and 3 reveal the secret to multiplying and squaring numbers in your head, quickly and accurately—the stock-in-trade of all "lightning calculators" (another name for mathemagicians). In Chapter 4, you will improve your ability to do mental division.

The next two chapters of the book move away from basic arithmetical functions and into other mental skills. Chapter 5 delves into the art of "guesstimation," and Chapter 6 shows you how to do rapid pencil-and-paper calculations. Among other things, I will show you how to write down the answer quickly to any 10-digit multiplication problem. In Chapter 7, on improving your mathematical memory, you will learn a method that will enable you to memorize the first 60 digits of pi (π) or any other number easily, in less than five minutes!

Chapter 8 makes the tough stuff—advanced multiplication—easy. It is mostly for hard-core mathemagicians, but anyone who makes it through the previous chapter and works hard in this one will not only be able to square a 5-digit number mentally but will also be able to multiply two different 5-digit numbers in his head—no pencil, no paper, no kidding!

Chapter 9, the last chapter, is my favorite. In fact, some of you may want to read it first since it does not depend on skills learned in the previous chapters. With very little practice you will learn to give the day of the week for any date in the past or future (move over Rain Man!), and other mathematical magic designed to make you look like a genius without really trying.

You will also find scattered throughout the book brief sidebars on mathematical geniuses and lightning calculators throughout history.

You may want to hunt around for things that interest you throughout this book. Flip through the pages and try your hand at a magic trick, or read one of the sidebars that jumps out at you. There is, to be sure, a sequence to the book, the easier stuff at the beginning, the more difficult toward the end. But many components of math are independent of each other and you may have fun just grabbing items here and there. And always, should you get bogged down at any point, you have only to go back a step or two to safer ground before diving back in. I can assure you that with practice you will get as much out of this book as you desire.

There is another subtle but more important benefit you will receive from this book: it will stimulate your creative thinking and show you the magic in math. Most people think of mathematics (and especially arithmetic) as mechanical, laborious, and rigid, with little room for creativity. The most pleasantly surprising lesson I believe you will discover in this book is that there are so many different ways to solve the same problem. The analogy with magic is fitting. Math is like magic. It's awe-inspiring to watch the numbers drop into place, and sometimes you can't believe your eyes.

I also want to emphasize at the beginning of this book that if you've ever seen a lightning calculator or a mental math whiz you probably got the impression that these people have some sort of special gift—that they are geniuses, or idiot savants, or people with unusual, genetically programmed ability. I can't speak for all lightning calculators, but in my case these skills were all learned. There is nothing I do that cannot be taught to someone else, even someone with no special aptitude for math beyond basic arithmetic. In short, it is my firmest belief that anyone can learn mathemagics if he or she is willing to invest some effort and practice.

ADVICE TO PARENTS AND TEACHERS

Wherever I perform I am asked by parents and teachers what they can do to motivate their children to excel in mathematics. For parents, my first piece of advice is, "Don't give up!" Let's face it, math can sometimes be hard, and does not come easily to many people. At some point, your children will come to you and say that they don't understand math. Whatever you do, do not say something like, "Oh well, I guess you're just not a math type." Mathematical literacy is too important a skill to be ignored in today's technological society. Yet many people actually brag about "never being good with figures." If your child had difficulty reading, would you just let it slide? Of course not. You would work with your child yourself, or hire a tutor or reading specialist, because you realize the value of being literate. Virtually everyone is capable of learning mathematics with enough hard work.

I have many strong feelings about how and what math should be taught, especially during a child's early years. We spend far too much time emphasizing long, laborious pencil-and-paper computations. Of course, children should be taught pencil-and-paper arithmetic, but once they are capable of working small problems by hand, enough is enough. Pencil-and-paper arithmetic is emphasized in schools way out of proportion to its practical value. Absolutely anyone needing the answer to a 3-digit times a 4-digit multiplication problem to seven digits of accuracy will have in his or her possession an inexpensive calculator. Who besides elementary school teachers and their students have to do long, long division problems by hand? Far more important is the ability to come up with an intelligent mental approximation of the final answer. Besides, when our students are compared with their international counterparts, Americans can do pencil-and-paper arithmetic with the best of them. Where we fail miserably is using mathematics to solve real-world problems. These dreaded "word problems" should be given far greater emphasis than we currently give them. With the wide availability of calculators (which I feel should be used in the classroom) it is more important for our students to know *when* mathematics is called for, and *what* numbers need to be calculated, than to master the mechanics of the calculation.

There are other benefits to learning mental arithmetic. Too often, mathematics is presented as a rigid set of rules, leaving little room for creativity. As you will learn from this book, there are often several different ways to solve the same problem. Large problems are broken down into smaller, more manageable components. We look for special features of the problem that we can exploit to make our job easier. These strike me as being valuable "life lessons" that we can use in approaching all kinds of problems, mathematical and otherwise. The process of taking a problem, solving it several different ways, and always winding up with the same numerical answer was my first exposure to the beauty and consistency of mathematics, and I've been hooked on math ever since. I hope that my enthusiasm will be contagious!

Arthur Benjamin
Claremont, California, 1993

1

A Little Give and Take: Mental Addition and Subtraction

I remember the day in third grade when I discovered that it was easier to add and subtract from left to right than from right to left, which was the way we had all been taught. Suddenly I was able to blurt out the answers to math problems in class well before my classmates put down their pencils. And I didn't even *need* a pencil! The method was so simple that I performed most calculations in my head. Looking back, I admit I did so as much to show off as for any mathematical reason. Most kids outgrow such behavior. Those who don't probably become either teachers or magicians.

In this chapter you will learn the left-to-right method of doing mental addition and subtraction for numbers that range in size from two to four digits. These mental skills are not only important for doing the tricks in this book but are also indispensable in school or at work, or any time you use numbers. Soon you will be able to retire your calculator and use the full capacity of your mind as you add, subtract, multiply, and divide 2-digit, 3-digit, and even 4-digit numbers.

LEFT-TO-RIGHT ADDITION

There are many good reasons why adding left to right is a superior method for *mental* calculation. For one thing, you do not have to reverse the numbers (as you do when adding right to left). And if you want to estimate your answer, then adding only the leading digits will get you pretty close. If you are used to working from right to left on paper, it may seem unnatural to add and multiply from left to right. But with practice you will find that it is the most natural and efficient way to do mental calculations.

With the first set of problems—2-digit addition—the left to right method may not seem so advantageous. But be patient. If you stick with me, you will see that the only *easy* way to solve 3-digit and larger addition problems, all subtraction problems, and most definitely all multiplication and division problems, is from left to right. The sooner you get accustomed to computing this way, the better.

2-DIGIT ADDITION

Our assumption in this chapter is that you know how to add and subtract 1-digit numbers. We will begin with 2-digit addition, something I suspect you can already do fairly well in your head. The following exercises are good practice, however, because you will use the 2-digit addition skills you polish here for larger addition problems, as well as in virtually all multiplication problems in later chapters. It also illustrates a fundamental principle of mental arithmetic—namely, to simplify your problem by breaking it into smaller, more manageable components. This is the key to virtually every method you will learn in this book. To paraphrase an old adage, there are just three components to success—simplify, simplify, simplify.

The easiest 2-digit addition problems, of course, are those that do not require you to carry any numbers. For example:

$$\begin{array}{r} 47 \\ + \ 32 \ (30 + 2) \\ \hline \end{array}$$

To add 32 to 47, you can simplify by treating 32 as 30 + 2, add 30 to 47 and then add 2. In this way the problem becomes 77 + 2, which equals 79:

$$\begin{array}{r} 47 \\ + \ 32 \end{array} \xrightarrow{+\ 30} \begin{array}{r} 77 \\ + \ 2 \end{array} \xrightarrow{+\ 2} = 79$$

Keep in mind that the above diagram is simply a way of representing the mental processes involved in arriving at an answer using one method. While you need to be able to read and understand such diagrams as you work your way through this book, our method does not require you to write down *anything* yourself.

Now let's try a calculation that requires you to carry a number:

$$67$$
$$+\ 28\ (20 + 8)$$

Adding from left to right, you can simplify the problem by adding $67 + 20 = 87$; then $87 + 8 = 95$.

$$67 \xrightarrow{+\ 20} \quad 87 \xrightarrow{+\ 8} = 95$$
$$+\ 28 \qquad\quad +\ 8$$

Now try one on your own, mentally calculating from left to right, and then check below to see how we did it:

$$84$$
$$+\ 57\ (50 + 7)$$

No problem, right? You added $84 + 50 = 134$ and added $134 + 7 = 141$.

$$84 \xrightarrow{+\ 50} \quad 134 \xrightarrow{+\ 7} = 141$$
$$+\ 57 \qquad\quad +\ 7$$

If carrying numbers trips you up a bit, don't worry about it. This is probably the first time you have ever made a systematic attempt at mental calculation, and if you're like most people, it will take you time to get used to it. With practice, however, you will begin to see and hear these numbers in your mind, and carrying numbers when you add will come automatically. Try another problem for practice, again computing it in your mind first, and then checking how we did it:

$$68$$
$$+\ 45\ (40 + 5)$$

You should have added $68 + 40 = 108$, and then $108 + 5 = 113$, the final answer. No sweat, right? If you would like to try your hand at more 2-digit addition problems, check out the set of exercises below. (The answers and computations are at the end of the book.)

Exercises: 2-Digit Addition

①
$$23$$
$$+ 16$$

②
$$64$$
$$+ 43$$

③
$$95$$
$$+ 32$$

④
$$34$$
$$+ 26$$

⑤
$$89$$
$$+ 78$$

⑥
$$73$$
$$+ 58$$

⑦
$$47$$
$$+ 36$$

⑧
$$19$$
$$+ 17$$

⑨
$$55$$
$$+ 49$$

⑩
$$39$$
$$+ 38$$

3-DIGIT ADDITION

The strategy for adding 3-digit numbers is the same as for adding 2-digit numbers: you add left to right. After each step, you arrive at a new (and smaller) addition problem. Let's try the following:

$$538$$
$$+ 327 \ (300 + 20 + 7)$$

After adding the hundreds digit of the second number to the first number (538 + 300 = 838), the problem becomes 838 + 27. Next add the tens digit (838 + 20 = 858), simplifying the problem to 858 + 7 = 865. This thought process can be diagrammed as follows:

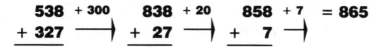

$$538 \ + 300 \qquad 838 \ + 20 \qquad 858 \ + 7 \qquad = 865$$
$$+ 327 \longrightarrow + 27 \longrightarrow + 7 \longrightarrow$$

All mental addition problems can be worked using this method. The goal is to keep simplifying the problem until you are left adding a 1-digit number. It is important to reduce the number of digits you are manipulating because human short-term memory is limited to about 7 digits. Notice that 538 + 327 requires you to hold on to 6 digits in your head, whereas 838 + 27 and 858 + 7 require only 5 and 4 digits, respectively. As you simplify the problems, the problems get easier!

Try the following addition problem in your mind before looking to see how we did it:

$$623$$
$$+ 159 \ (100 + 50 + 9)$$

Did you reduce and simplify the problem by adding left to right? After adding the hundreds digit (623 + 100 = 723), you were left with 723 + 59. Next you should have added the tens digit (723 + 50 = 773), simplifying the problem to 773 + 9, which you easily summed to 782. Diagrammed, the problem looks like this:

$$\begin{array}{r} 623 \\ + \ 159 \end{array} \xrightarrow{\ +\ 100\ } \begin{array}{r} 723 \\ + \ 59 \end{array} \xrightarrow{\ +\ 50\ } \begin{array}{r} 773 \\ + \ \ \ 9 \end{array} \xrightarrow{\ +\ 9\ } = 782$$

When I do these problems mentally, I do not try to *see* the numbers in my mind—I try to *hear* them. I hear the problem 623 + 159 as six **hundred** twenty-three plus one **hundred** fifty-nine; by emphasizing the word "hundred" to myself, I know where to begin adding. Six plus one equals seven, so my next problem is *seven* **hundred** *and twenty-three* plus *fifty-nine*, and so on. When first doing these problems, practice them out loud. Reinforcing yourself verbally will help you learn the mental method much more quickly.

Addition problems really do not get much harder than the following:

$$\begin{array}{r} 858 \\ + \ 634 \end{array}$$

Now look to see how we did it, below:

$$\begin{array}{r} 858 \\ + \ 634 \end{array} \xrightarrow{\ +\ 600\ } \begin{array}{r} 1458 \\ + \ \ \ 34 \end{array} \xrightarrow{\ +\ 30\ } \begin{array}{r} 1488 \\ + \ \ \ \ 4 \end{array} \xrightarrow{\ +\ 4\ } = 1492$$

At each step I hear (not see) a "new" addition problem. In my mind the problem sounds like this:

858 plus 634 is 1458 plus 34 is 1488 plus 4 is 1492

Your mind-talk may not sound exactly like mine, but whatever it is you say to yourself, the point is to reinforce the numbers along the way so that you don't forget where you are and have to start the addition problem over again.

Let's try another one for practice:

$$759$$
$$+ \ 496 \ (400 + 90 + 6)$$

Do it in your mind first, then check our computation, below:

$$\begin{array}{r} 759 \\ + \ 496 \end{array} \xrightarrow{+ \ 400} \begin{array}{r} 1159 \\ + \ \ 96 \end{array} \xrightarrow{+ \ 90} \begin{array}{r} 1249 \\ + \ \ \ \ 6 \end{array} \xrightarrow{+ \ 6} = 1255$$

This addition problem is a little more difficult than the last one since it requires you to carry numbers in all three steps. However, with this particular problem you have the option of using an alternative method. I am sure you will agree that it is a lot easier to add 500 to 759 than it is to add 496, so try adding 500 and then subtracting the difference:

$$759$$
$$+ \ 496 \ (500 - 4)$$

$$\begin{array}{r} 759 \\ + \ 496 \end{array} \xrightarrow{+ \ 500} \begin{array}{r} 1259 \\ - \ \ \ \ 4 \end{array} \xrightarrow{- \ 4} = 1255$$

So far, you have consistently broken up the *second* number in any problem to add to the first. It really does not matter which number you choose to break up as long as you are consistent. That way, your mind will never have to waste time deciding which way to go. If the second number happens to be a lot simpler than the first, I switch them around, as in the following example:

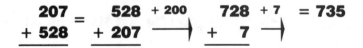

$$\begin{array}{r} 207 \\ + \ 528 \end{array} = \begin{array}{r} 528 \\ + \ 207 \end{array} \xrightarrow{+ \ 200} \begin{array}{r} 728 \\ + \ \ \ \ 7 \end{array} \xrightarrow{+ \ 7} = 735$$

Let's finish up by adding 3-digit to 4-digit numbers. Again, since most human memory can only hold about seven digits at a time, this is about as large a problem as you can handle without resorting to artificial memory devices (described in Chapter 7). Often (especially within multiplication problems) one or both of the numbers will end in 0, so we shall emphasize those types of problems. We begin with an easy one:

$$\begin{array}{r} 2700 \\ + \ 567 \\ \hline \end{array}$$

Since 27 hundred + 5 hundred is 32 hundred, we simply attach the 67 to get 32 hundred and 67, or 3267. The process is the same for the following problems:

$$\begin{array}{r} 3240 \\ + \ \ 18 \\ \hline \end{array} \qquad \begin{array}{r} 3240 \\ + \ \ 72 \\ \hline \end{array}$$

Because 40 + 18 = 58, the first answer is 3258. For the second problem, since 40 + 72 exceeds 100, you know the answer will be 33 hundred and *something*. Because 40 + 72 = 112, you end up with 3312.

These problems are easy because the digits only overlap in one place, and hence can be solved in a single step. Where digits overlap in two places, you require two steps. For instance:

$$\begin{array}{r} 4560 \\ + \ 171 \ (100 + 71) \\ \hline \end{array}$$

This problem requires two steps, as diagrammed the following way:

$$\begin{array}{r} 4560 \\ + \ 171 \\ \hline \end{array} \xrightarrow{+\ 100} \begin{array}{r} 4660 \\ + \ \ 71 \\ \hline \end{array} \xrightarrow{+\ 71} = 4731$$

Practice the following 3-digit addition exercises, and then add some of your own if you like (pun intended!) until you are comfortable doing them mentally without having to look down at the page.

Carl Friedrich Gauss: Mathematical Prodigy

A prodigy is a highly talented child, usually called precocious, or gifted, and almost always ahead of his peers. The German mathematician Carl Friedrich Gauss (1777–1855) was one such child. He often boasted that he could calculate before he could speak. By the ripe old age of three, before he had been taught any arithmetic, he corrected his father's payroll by declaring "the reckoning is wrong." A further check of the numbers proved young Carl correct.

As a ten-year-old student, Gauss was presented the following mathematical problem: What is the sum of numbers from 1 to 100? While his fellow students were frantically calculating with paper and pencil, Gauss immediately envisioned that if he spread out the numbers 1 through 50 from left to right, and the numbers 51 through 100 from right to left directly below the 1–50 numbers, each combination would add to 101 (1 + 100; 2 + 99; 3 + 98 . . .). Since there were 50 combinations, the answer would be 101 × 50 = 5050. To the astonishment of everyone, including the teacher, young Carl got the answer not only ahead of everyone else, but computed it entirely in his mind. He wrote out the answer on his slate, and flung it on the teacher's desk with a defiant: "There it lies." The teacher was so impressed that he invested his own money to purchase the best available textbook on arithmetic and gave it to Gauss, stating: "He is beyond me, I can teach him nothing more."

Indeed, Gauss became the mathematics teacher of others, and eventually went on to become one of the greatest mathematicians in history, his theories still used today in the service of science. Gauss' desire to better understand nature through the language of mathematics was summed up in his motto, taken from Shakespeare's *King Lear* (substituting "laws" for "law"): "Thou, nature, art my goddess; to thy laws My services are bound."

Exercises: 3-Digit Addition

①
$$242$$
$$+ 137$$

②
$$312$$
$$+ 256$$

③
$$635$$
$$+ 814$$

④
$$457$$
$$+ 241$$

⑤
$$912$$
$$+ 475$$

⑥
$$852$$
$$+ 378$$

⑦
$$457$$
$$+ 269$$

⑧
$$878$$
$$+ 797$$

⑨
$$276$$
$$+ 689$$

⑩
$$877$$
$$+ 539$$

⑪	⑫	⑬	⑭	⑮
5400	1800	6120	7830	4240
+ 252	+ 855	+ 136	+ 348	+ 371

Answers can be found in the back of the book.

LEFT-TO-RIGHT SUBTRACTION

For most of us, it is easier to add than to subtract. But if you continue to compute from left to right and to break problems down into simpler components (using the principle of simplification, as always), subtraction can become almost as easy as addition.

2-DIGIT SUBTRACTION

For 2-digit subtraction, as in addition, your goal is to keep simplifying the problem until you are reduced to subtracting a 1-digit number. Let's begin with a very simple subtraction problem:

$$86$$
$$- 25 \ (20 + 5)$$

After each step, you arrive at a new and smaller subtraction problem. Begin by subtracting: $86 - 20 = 66$. Your problem becomes $66 - 5 = 61$, your final answer. The problem can be diagrammed this way:

$$\begin{array}{ccc} 86 & -20 \\ -25 & \longrightarrow \end{array} \quad \begin{array}{cc} 66 & -5 \\ -5 & \longrightarrow \end{array} = 61$$

Of course, subtraction problems are considerably easier when they do not involve borrowing. When they do, there are a number of strategies you can use to make them easier. For example:

$$86$$
$$- 29 \ (20 + 9) \text{ or } (30 - 1)$$

There are two different ways to solve this problem mentally:

1. You can simplify the problem by breaking down 29 into 20 and 9, subtracting 20, then subtracting 9:

$$(1) \quad \begin{array}{r} 86 \\ -\ 29 \\ \hline \end{array} \overset{-\ 20}{\longrightarrow} \quad \begin{array}{r} 66 \\ -\ 9 \\ \hline \end{array} \overset{-\ 9}{\longrightarrow} = 57$$

2. You can treat 29 as 30 − 1, subtracting 30, then adding back 1:

$$(2) \quad \begin{array}{r} 86 \\ -\ 29 \\ \hline \end{array} \overset{-\ 30}{\longrightarrow} \quad \begin{array}{r} 56 \\ +\ 1 \\ \hline \end{array} \overset{+\ 1}{\longrightarrow} = 57$$

Here is the rule for deciding which method to use:

If the subtraction of two numbers requires you to borrow a number, round up the second number to a multiple of ten and add back the difference.

Here is another example of rounding up:

$$\begin{array}{r} 54 \\ -\ 28\ (30 - 2) \\ \hline \end{array}$$

Since this problem requires you to borrow, round up 28 to 30, subtract, and then add back 2 to get 26 as your final answer:

$$\begin{array}{r} 54 \\ -\ 28 \\ \hline \end{array} \overset{-\ 30}{\longrightarrow} \quad \begin{array}{r} 24 \\ +\ 2 \\ \hline \end{array} \overset{+\ 2}{\longrightarrow} = 26$$

Now try your hand at this 2-digit subtraction problem:

$$\begin{array}{r} 81 \\ -\ 37 \\ \hline \end{array}$$

Easy, right? You just round up 37 to 40, subtract 40 from 81, which gives you 41, and then add back the difference of 3 to arrive at 44, the final answer:

$$\begin{array}{r} 81 \\ -\ 37 \\ \hline \end{array} \overset{-\ 40}{\longrightarrow} \quad \begin{array}{r} 41 \\ +\ 3 \\ \hline \end{array} \overset{+\ 3}{\longrightarrow} = 44$$

In a short time you will become comfortable working subtraction problems both ways. Just use the rule above to decide which method will work best.

Exercises: 2-Digit Subtraction

①
$$38 - 23$$

②
$$84 - 59$$

③
$$92 - 34$$

④
$$67 - 48$$

⑤
$$79 - 29$$

⑥
$$63 - 46$$

⑦
$$51 - 27$$

⑧
$$89 - 48$$

⑨
$$125 - 79$$

⑩
$$148 - 86$$

3-DIGIT SUBTRACTION

Now let's try a 3-digit subtraction problem:

$$958 - 417 \ (400 + 10 + 7)$$

This particular problem does not require you to borrow any numbers, so you should not find it *too* hard. Simply subtract one digit at a time, simplifying as you go.

$$\underset{-\ 417}{958} \xrightarrow{-400} \underset{-\ 17}{558} \xrightarrow{-10} \underset{-\ 7}{548} \xrightarrow{-7} = 541$$

Now let's look at a 3-digit subtraction problem that requires you to borrow a number:

$$747 - 598 \ (600 - 2)$$

At first glance this probably looks like a pretty tough problem, but if you round up by 2, subtract $747 - 600 = 147$, then add back 2, you reach your final answer of $147 + 2 = 149$.

$$\underset{-\ 598}{747} \xrightarrow{-600} \underset{+\ 2}{147} \xrightarrow{+2} = 149$$

Now try one yourself:

$$853$$
$$- 692$$

Did you round 692 up to 700 and then subtract 700 from 853? If you did, you got 853 − 700 = 153. Since you subtracted by 8 too much, did you add back 8 to reach 161, the final answer?

$$853 \quad -700 \qquad 153 \quad +8 \quad = 161$$
$$- 692 \longrightarrow \quad + \quad 8 \longrightarrow$$

Now, I admit we have been making life easier for you by choosing numbers you don't have to round up by much. But what happens when it isn't so easy to figure out how much to add back when you have subtracted too much? The following 3-digit subtraction problem illustrates exactly what I mean:

$$725$$
$$- 468 \ (400 + 60 + 8) \text{ or } (500 - \text{??})$$

If you subtract one digit at a time, simplifying as you go, your sequence will look like this:

$$725 \quad -400 \qquad 325 \quad -60 \qquad 265 \quad -8 \quad = 257$$
$$- 468 \longrightarrow \quad - 68 \longrightarrow \quad - 8 \longrightarrow$$

What happens if you round up to 500?

$$725 \quad -500 \quad = \quad 225$$
$$- 468 \longrightarrow \quad + \quad \text{??}$$

Subtracting 500 is easy: 725 − 500 = 225. But you have subtracted too much. The trick is to figure out exactly how much.

At first glance, the answer is far from obvious. To find it, you need to know how far 468 is from 500. The answer can be found by using "complements," a nifty technique that will make many 3-digit subtraction problems a lot easier to figure out.

USING COMPLEMENTS (YOU'RE WELCOME!)

Quick, how far from 100 is each of these numbers?

57   **68**   **49**   **21**   **79**

Here are the answers:

57	68	49	21	79
+ 43	+ 32	+ 51	+ 79	+ 21
100	**100**	**100**	**100**	**100**

Notice that for each pair of numbers that add to 100, the first digits (on the left) add to 9 and the last (on the right) add to 10. We say that 43 is the complement of 57, 32 the complement of 68, and so on.

Now *you* find the complement of these 2-digit numbers:

37   **59**   **93**   **44**   **08**

To find the complement of 37, first figure out what you need to add to 3 to get 9. (The answer is 6.) Then figure out what you need to add to 7 to get 10. (The answer is 3.) Hence, 63 is your complement.

The other complements are 41, 7, 56, 92. Notice that, like everything else you do as a mathemagician, the complements are determined from left to right. As we have seen, the first digits add to 9, and the second digits add to 10. (An exception occurs in numbers ending in 0—e.g., 30 + 70 = 100—but those complements are simple!)

What do complements have to do with mental subtraction? Well, they allow you to convert difficult subtraction problems into straightforward addition problems. Let's consider the last subtraction problem that gave us some trouble:

$$\begin{array}{r} 725 \\ -\ 468\ (500 - 32) \\ \hline \end{array}$$

To begin, you subtracted 500 instead of 468 to arrive at 225 (725 − 500 = 225). But then, having subtracted too much, you needed to figure out how much too much. Using complements gives you the answer in a flash. How far is 468 from 500? The same distance as 68 is from 100. If you take the complement of 68 the way we have shown you, you will arrive at 32. Add 32 to 225 and you will arrive at 257, your final answer.

$$725 \xrightarrow{-500} 225 \xrightarrow{+32} = 257$$
$$-468 \qquad +32$$

Try another 3-digit subtraction problem:

$$821$$
$$- 259 \ (300 - 41)$$

To compute this mentally, subtract 300 from 821 to arrive at 521, then add back the complement of 59, which is 41, to arrive at 562, our final answer. The procedure looks like this:

$$821 \xrightarrow{-300} 521 \xrightarrow{+41} = 562$$
$$-259 \qquad +41$$

Here is another problem for you to try:

$$645$$
$$- 372 \ (400 - 28)$$

Check your answer and the procedure for solving the problem, below:

$$645 \xrightarrow{-400} 245 \xrightarrow{+20} 265 \xrightarrow{+8} = 273$$
$$-372 \qquad +28 \qquad +8$$

Subtracting a 3-digit number from a 4-digit number is not much harder, as the next example illustrates:

$$1246$$
$$- 579 \ (600 - 21)$$

By rounding up you subtract 600 from 1246, leaving 646, then add back the complement of 79, which is 21. Your final answer is 646 + 21 = 667.

$$1246 \xrightarrow{-600} 646 \xrightarrow{+21} = 667$$
$$- 579 \qquad\qquad + 21$$

We will have more to say about subtraction in Chapters 3 and 8. In the meantime, try the 3-digit subtraction exercises below, and then create more of your own for additional (or should that be subtractional?) practice.

Exercises: 3-Digit Subtraction

①
$$583$$
$$- 271$$

②
$$936$$
$$- 725$$

③
$$587$$
$$- 298$$

④
$$763$$
$$- 486$$

⑤
$$204$$
$$- 185$$

⑥
$$793$$
$$- 402$$

⑦
$$219$$
$$- 176$$

⑧
$$978$$
$$- 784$$

⑨
$$455$$
$$- 319$$

⑩
$$772$$
$$- 596$$

⑪
$$873$$
$$- 357$$

⑫
$$564$$
$$- 228$$

⑬
$$1428$$
$$- 571$$

⑭
$$2345$$
$$- 678$$

⑮
$$1776$$
$$- 987$$

Products of a Misspent Youth: Basic Multiplication

I spent practically my entire childhood devising faster and faster ways to perform mental multiplication; hence the title of this chapter. In hindsight I suppose my youth wasn't really misspent, since my classroom antics led me straight into a career in mathematics. But I know that at the time plenty of my teachers were frustrated with me. I was diagnosed as hyperactive and my parents were told that I had a short attention span and probably would not be successful in school. Ironically, it was my short attention span that motivated me to develop quick ways to do arithmetic. I could not possibly sit still long enough to carry out math problems with pencil and paper. Once you have mastered the techniques described in this chapter, you won't want to rely on pencil and paper again, either.

In this chapter you will learn how to multiply 1-digit numbers by 2-digit numbers and 3-digit numbers in your head, which will lay the foundation for the more complicated multiplication problems to come. You will also learn a phenomenally fast way to square 2-digit numbers. Even friends with calculators won't be able to keep up with you. Believe me, virtually everyone will be dumbfounded by the fact that such problems can not only be done mentally, but can be computed so quickly. I sometimes wonder whether we were not cheated in school; these methods are so simple once you learn them.

There *is* one small prerequisite for mastering the mathemagic tricks in this chapter—you need to know the multiplication tables through 10. In fact, to really make headway, you need to know your multiplication tables backward and forward. For those of you who need to shake the cobwebs loose, consult the illustration heading for this chapter. Once you've got your tables down, you are in for some fun, because multiplication provides ample opportunities for creative problem solving.

2-BY-1 MULTIPLICATION PROBLEMS

If you worked your way through the last chapter, you got into the habit of adding and subtracting from left to right. You will do virtually all the calculations in this chapter from left to right, as well. This is undoubtedly the opposite of what you learned in school. But you'll soon see how much easier it is to think from left to right than from right to left. (For one thing, you can start to say your answer aloud before you have finished the calculation. That way you *seem* to be calculating even faster than you are!)

Let's tackle our first problem:

$$\begin{array}{r} 42 \\ \times\ \ 7 \\ \hline \end{array}$$

First, multiply $40 \times 7 = 280$. Next, multiply $2 \times 7 = 14$. Add 14 to 280 (left to right, of course) to arrive at 294, the correct answer. We illustrate this procedure below. We have omitted diagramming the mental addition of $280 + 14$, since you learned in Chapter 1 how to do this computation.

$$\begin{array}{rr} & 42\ (40 + 2) \\ \times & 7 \\ \hline 40 \times 7 = & 280 \\ 2 \times 7 = & +\ 14 \\ \hline & 294 \end{array}$$

At first you will need to look down at the problem while calculating it to recall the next operation. With practice you will be able to forgo this step and compute the whole thing in your mind.

Let's try another example:

$$
\begin{array}{r}
48 \\
\times\ 4 \\
\hline
\end{array}
$$

Your first step is to break down the problem into small multiplication tasks that you can perform mentally with ease. Since 48 = 40 + 8, multiply 40 × 4 = 160, then add 8 × 4 = 32. The answer is 192. (Note: If you are wondering *why* this process works, see the section "Why These Tricks Work" at the end of the chapter.)

$$
\begin{array}{r}
48\ (40 + 8) \\
\times\quad 4 \\
\hline
\end{array}
$$

$$
\begin{array}{rr}
40 \times 4 = & 160 \\
8 \times 4 = & +\ 32 \\
\hline
& 192
\end{array}
$$

Here are two more mental multiplication problems that you should be able to get the answers to fairly quickly. Try calculating them in your head before looking at how we did it.

$$
\begin{array}{rr}
62\ (60 + 2) \\
\times\quad 3 \\
\hline
60 \times 3 = & 180 \\
2 \times 3 = +\ & 6 \\
\hline
& 186
\end{array}
\qquad
\begin{array}{rr}
71\ (70 + 1) \\
\times\quad 9 \\
\hline
70 \times 9 = & 630 \\
1 \times 9 = +\ & 9 \\
\hline
& 639
\end{array}
$$

These two examples are especially simple because they do not require you to carry any numbers. Another especially easy type of mental multiplication problem involves numbers that begin with 5. When the 5 is multiplied by an even digit, the first product will be a multiple of 100, which makes the resulting addition problem a snap:

$$
\begin{array}{r}
58\ (50 + 8) \\
\times\quad 4 \\
\hline
\end{array}
$$

$$
\begin{array}{rr}
50 \times 4 = & 200 \\
8 \times 4 = & +\ 32 \\
\hline
& 232
\end{array}
$$

Try your hand at the following problem:

$$
\begin{array}{r}
\textbf{87 (80 + 7)} \\
\times \quad \textbf{5} \\
\hline
80 \times 5 = \quad \textbf{400} \\
7 \times 5 = + \ \textbf{35} \\
\hline
\textbf{435}
\end{array}
$$

Notice how much easier this problem is to do from left to right. It takes far less time to calculate "400 plus 35" mentally than it does to apply the pencil-and-paper method of "putting down the 5 and carrying the 3."

The following two problems are harder because they force you to carry numbers when you come to the addition:

$$
\begin{array}{r}
\textbf{38 (30 + 8)} \\
\times \quad \textbf{9} \\
\hline
30 \times 9 = \quad \textbf{270} \\
8 \times 9 = + \ \textbf{72} \\
\hline
\textbf{342}
\end{array}
\qquad
\begin{array}{r}
\textbf{67 (60 + 7)} \\
\times \quad \textbf{8} \\
\hline
60 \times 8 = \quad \textbf{480} \\
7 \times 8 = + \ \textbf{56} \\
\hline
\textbf{536}
\end{array}
$$

As usual, break these problems down into easier problems. For the one on the left, multiply 30×9 plus 8×9, giving you $270 + 72$. The addition problem is slightly harder because it involves carrying a number. Here $270 + 70 + 2 = 340 + 2 = 342$.

With practice, you will become more adept at juggling problems like these in your head, and those that require you to carry numbers will be almost as easy as the others.

Rounding Up

You saw in the last chapter how useful rounding up can be when it comes to subtraction. The same goes for multiplication, especially when the numbers you are multiplying end in an 8 or 9.

Let's take the problem of 69×6, illustrated below. On the left we have calculated it the usual way, by adding. On the right, however, we have rounded 69 up to 70, which is an easy number to multiply. And for many people, it is easier to subtract $420 - 6$ than it is to add $360 + 54$ when calculating mentally.

$$\begin{array}{r} \textbf{69 (60 + 9)} \\ \times \quad \textbf{6} \\ \hline \end{array} \qquad \textbf{or} \qquad \begin{array}{r} \textbf{69 (70 - 1)} \\ \times \quad \textbf{6} \\ \hline \end{array}$$

$$\begin{array}{rr} \textbf{60} \times \textbf{6} = & \textbf{360} \\ \textbf{9} \times \textbf{6} = & +\ \textbf{54} \\ \hline & \textbf{414} \end{array} \qquad \begin{array}{rr} \textbf{70} \times \textbf{6} = & \textbf{420} \\ -\ \textbf{1} \times \textbf{6} = & -\quad \textbf{6} \\ \hline & \textbf{414} \end{array}$$

The following example also shows how much easier rounding up can be:

$$\begin{array}{r} \textbf{78 (70 + 8)} \\ \times \quad \textbf{9} \\ \hline \end{array} \qquad \textbf{or} \qquad \begin{array}{r} \textbf{78 (80 - 2)} \\ \times \quad \textbf{9} \\ \hline \end{array}$$

$$\begin{array}{rr} \textbf{70} \times \textbf{9} = & \textbf{630} \\ \textbf{8} \times \textbf{9} = & +\ \textbf{72} \\ \hline & \textbf{702} \end{array} \qquad \begin{array}{rr} \textbf{80} \times \textbf{9} = & \textbf{720} \\ -\ \textbf{2} \times \textbf{9} = & -\ \textbf{18} \\ \hline & \textbf{702} \end{array}$$

The subtraction method works especially well for numbers just one or two digits away from a multiple of 10. It does not work so well when you need to round up more than two digits because the subtraction portion of the problem gets out of hand. As it is, you may prefer to stick with the addition method. Personally, I only use the addition method because in the time spent deciding which method to use, I could have already done the calculation!

So that you can perfect your technique we *strongly* recommend practicing more 2-by-1 multiplication problems. Below are 20 problems for you to tackle. We have supplied you with the answers in the back of the book, including a breakdown of each component of the multiplication. If, after you've worked out these problems, you would like to practice more, make up your own. Calculate mentally, then check your answer with a calculator. Once you feel confident that you can perform these problems rapidly in your head, you are ready to move to the next level of mental calculation.

Exercises: 2-by-1 Multiplication

①	②	③	④	⑤
82	43	67	71	93
× 9	× 7	× 5	× 3	× 8

⑥	⑦	⑧	⑨	⑩
49	28	53	84	58
× 9	× 4	× 5	× 5	× 6

⑪
97
× 4
———

⑫
78
× 2
———

⑬
96
× 9
———

⑭
75
× 4
———

⑮
57
× 7
———

⑯
37
× 6
———

⑰
46
× 2
———

⑱
76
× 8
———

⑲
29
× 3
———

⑳
64
× 8
———

3-BY-1 MULTIPLICATION PROBLEMS

Now that you know how to do 2-by-1 multiplication problems in your head, you will find that multiplying three digits by a single digit is not much more difficult. You can get started with the following 3-by-1 problem (which is really just a 2-by-1 problem in disguise):

$$
\begin{array}{r}
320 \ (300 + 20) \\
\times \qquad 7 \\
\hline
300 \times 7 = \qquad 2100 \\
20 \times 7 = + \quad 140 \\
\hline
2240
\end{array}
$$

What could be easier? Let's try another 3-by-1 problem similar to the one you just did, except we have replaced the 0 with a 6 so that you have another step to perform:

$$
\begin{array}{r}
326 \ (300 + 20 + 6) \\
\times \qquad 7 \\
\hline
300 \times 7 = \qquad 2100 \\
20 \times 7 = + \quad 140 \\
\hline
2240 \\
6 \times 7 = + \quad 42 \\
\hline
2282
\end{array}
$$

In this case, you simply add the product of 6 × 7, which you already know to be 42, to the first sum of 2240. Since you do not need to carry any numbers, it is easy to add 42 to 2240 to arrive at the total of 2282.

In solving this and other 3-by-1 multiplication problems, the difficult part may be holding in memory the first sum (in this case, 2240) while doing the next multiplication problem (in this case, 6 × 7). There is no magic secret to remembering that first number, but with practice I guarantee you will improve your concentration so that holding on to numbers while performing other functions will get easier.

Let's try another problem:

$$
\begin{array}{r}
\textbf{647 (600 + 40 + 7)} \\
\times\quad\ \textbf{4} \\
\hline
\end{array}
$$

$$
\begin{array}{rr}
\textbf{600} \times \textbf{4} = & \textbf{2400} \\
\textbf{40} \times \textbf{4} = & +\ \textbf{160} \\
\hline
& \textbf{2560} \\
\textbf{7} \times \textbf{4} = & +\ \ \ \textbf{28} \\
\hline
& \textbf{2588}
\end{array}
$$

Even if the numbers are large, the process is just as simple. For example:

$$
\begin{array}{r}
\textbf{987 (900 + 80 + 7)} \\
\times\quad\ \textbf{9} \\
\hline
\end{array}
$$

$$
\begin{array}{rr}
\textbf{900} \times \textbf{9} = & \textbf{8100} \\
\textbf{80} \times \textbf{9} = & +\ \textbf{720} \\
\hline
& \textbf{8820} \\
\textbf{7} \times \textbf{9} = & +\ \ \ \textbf{63} \\
\hline
& \textbf{8883}
\end{array}
$$

When first solving these problems, you may have to glance down at the page as you go along to remind yourself what the original problem was. This is okay at first. But try to break the habit so that eventually you are holding the problem entirely in memory.

In the last section on 2-by-1 multiplication problems, we saw that problems involving numbers that begin with 5 are sometimes especially easy to solve. The same is true for 3-by-1 problems:

$$
\begin{array}{r}
\textbf{563 (500 + 60 + 3)} \\
\times \quad \textbf{6} \\
\hline
\textbf{500} \times \textbf{6} = \quad \textbf{3000} \\
\textbf{60} \times \textbf{6} = \quad \textbf{360} \\
\textbf{3} \times \textbf{6} = \quad \textbf{18} \\
\hline
\textbf{3378}
\end{array}
$$

Notice that whenever the first product is a multiple of 1000, the resulting addition problem is no problem at all because you do not have to carry any numbers and the thousands digit does not change. If you were solving the problem above in front of an audience, you would be able to say your first product—"3000 . . ."—out loud with complete confidence that a carried number would not change it to 4000. (As an added bonus, by quickly saying the first digit, it gives the illusion that you computed the answer immediately!) Even if you are practicing alone, saying your first product out loud frees up some memory space while you work on the remaining 2-by-1 problem, which you can say out loud as well—in this case, ". . . three hundred seventy-eight."

Try the same approach in solving the next problem, where the multiplier is a 5:

$$
\begin{array}{r}
\textbf{663 (600 + 60 + 3)} \\
\times \quad \textbf{5} \\
\hline
\textbf{600} \times \textbf{5} = \quad \textbf{3000} \\
\textbf{60} \times \textbf{5} = \quad \textbf{300} \\
\textbf{3} \times \textbf{5} = + \quad \textbf{15} \\
\hline
\textbf{3315}
\end{array}
$$

Because the first *two* digits of the 3-digit number are even, you can say the answer as you calculate it without having to add anything! Don't you wish all multiplication problems were so easy?

Let's escalate the challenge by trying a couple of problems that require you to carry a number:

$$
\begin{array}{r}
184\ (100 + 80 + 4) \\
\times \qquad 7 \\
\hline
\end{array}
$$

$$
\begin{array}{r}
100 \times 7 = \quad 700 \\
80 \times 7 = +\ 560 \\
\hline
1260 \\
4 \times 7 = +\quad 28 \\
\hline
1288
\end{array}
$$

$$
\begin{array}{r}
684\ (600 + 80 + 4) \\
\times \qquad 9 \\
\hline
\end{array}
$$

$$
\begin{array}{r}
600 \times 9 = \quad 5400 \\
80 \times 9 = +\ 720 \\
\hline
6120 \\
4 \times 9 = +\quad 36 \\
\hline
6156
\end{array}
$$

In the next two problems you need to carry a number at the end of the problem instead of at the beginning:

$$
\begin{array}{r}
648\ (600 + 40 + 8) \\
\times \qquad 9 \\
\hline
\end{array}
$$

$$
\begin{array}{r}
600 \times 9 = \quad 5400 \\
40 \times 9 = +\ 360 \\
\hline
5760 \\
8 \times 9 = +\quad 72 \\
\hline
5832
\end{array}
$$

$$
\begin{array}{r}
\textbf{376 (300 + 70 + 6)} \\
\times \quad\ \ \textbf{4} \\
\hline
\end{array}
$$

$$
\begin{array}{rr}
\textbf{300} \times \textbf{4} = & \textbf{1200} \\
\textbf{70} \times \textbf{4} = & +\ \ \textbf{280} \\
\hline
& \textbf{1480} \\
\textbf{6} \times \textbf{4} = & +\ \ \ \ \textbf{24} \\
\hline
& \textbf{1504}
\end{array}
$$

The first part of each of these problems is easy enough to compute mentally. The difficult part comes in holding the preliminary answer in your head while computing the final answer. In the case of the first problem, it is easy to add 5400 + 360 = 5760, but you may have to repeat 5760 to yourself several times while you multiply 8 × 9 = 72. Then add 5760 + 72. Sometimes at this stage I will start to say my answer aloud before finishing. Because I know I will have to carry when I add 60 + 72, I know that 5700 will become 5800, so I say "fifty-eight hundred . . ." Then I pause to compute 60 + 72 = 132. Because I have already carried, I say only the last two digits, ". . . thirty-two!" And there is the answer: 5832.

The next two problems require you to carry two numbers each, so they may take you longer than those you have already done. But with practice you will get faster:

$$
\begin{array}{r}
\textbf{489 (400 + 80 + 9)} \\
\times \quad\ \ \textbf{7} \\
\hline
\end{array}
$$

$$
\begin{array}{rr}
\textbf{400} \times \textbf{7} = & \textbf{2800} \\
\textbf{80} \times \textbf{7} = & +\ \ \textbf{560} \\
\hline
& \textbf{3360} \\
\textbf{9} \times \textbf{7} = & +\ \ \ \ \textbf{63} \\
\hline
& \textbf{3423}
\end{array}
$$

$$
\begin{array}{r}
\textbf{224 (200 + 20 + 4)} \\
\times \quad \textbf{9} \\
\hline
\end{array}
$$

$$
\begin{array}{rr}
\textbf{200} \times \textbf{9} = & \textbf{1800} \\
\textbf{20} \times \textbf{9} = & + \ \textbf{180} \\
\hline
& \textbf{1980} \\
\textbf{4} \times \textbf{9} = & + \quad \textbf{36} \\
\hline
& \textbf{2016}
\end{array}
$$

When you are first tackling these problems, repeat the answers to each part out loud as you compute the rest. In the first problem, for example, start out by saying, "Twenty-eight hundred plus five hundred and sixty" a couple of times out loud to reinforce the two numbers in memory while you add them together. Repeat the answer—"thirty-three hundred and sixty"—several times while you multiply $9 \times 7 = 63$. Then repeat "thirty-three hundred and sixty plus sixty-three" aloud until you compute the final answer of 3423. If you are thinking fast enough to recognize that adding $60 + 63$ will require you to carry a 1, you can begin to give the final answer a split second before you know it—"thirty-four thousand . . . and twenty-three!"

Let's end this section on 3-by-1 multiplication problems with some special problems you can do in a flash because they require one addition step instead of two:

$$
\begin{array}{r}
\textbf{511 (500 + 11)} \\
\times \quad \textbf{7} \\
\hline
\end{array}
$$

$$
\begin{array}{rr}
\textbf{500} \times \textbf{7} = & \textbf{3500} \\
\textbf{11} \times \textbf{7} = & + \quad \textbf{77} \\
\hline
& \textbf{3577}
\end{array}
$$

$$
\begin{array}{r}
\textbf{925 (900 + 25)} \\
\times \quad \textbf{8} \\
\hline
\end{array}
$$

$$
\begin{array}{rr}
\textbf{900} \times \textbf{8} = & \textbf{7200} \\
\textbf{25} \times \textbf{8} = & + \ \textbf{200} \\
\hline
& \textbf{7400}
\end{array}
$$

$$825 \ (800 + 25)$$
$$\times \quad 3$$
$$800 \times 3 = \quad 2400$$
$$25 \times 3 = + \quad 75$$
$$\overline{\qquad 2475}$$

In general, if the product of the last two digits of the first number and the multiplier is known to you without having to calculate it (for instance, you may know that $25 \times 8 = 200$ automatically since 8 quarters equals $2.00), you will get to the final answer much more quickly. For instance, *if* you know without calculating that $75 \times 4 = 300$, then it is a breeze to compute 975×4:

$$975 \ (900 + 75)$$
$$\times \quad 4$$
$$900 \times 4 = \quad 3600$$
$$75 \times 4 = \quad 300$$
$$\overline{\qquad 3900}$$

To reinforce what you have just learned, solve the following 3-by-1 multiplication problems in your head; then check your computations and answers with ours (in the back of the book). I can assure you from experience that doing mental calculations is just like riding a bicycle or typing. It might seem impossible at first, but once you've mastered it, you will never forget how to do it.

Exercises: 3-by-1 Multiplication

①
$$431$$
$$\times \quad 6$$

②
$$637$$
$$\times \quad 5$$

③
$$862$$
$$\times \quad 4$$

④
$$957$$
$$\times \quad 6$$

⑤
$$927$$
$$\times \quad 7$$

⑥
$$728$$
$$\times \quad 2$$

⑦
$$328$$
$$\times \quad 6$$

⑧
$$529$$
$$\times \quad 9$$

⑨ 807 × 9   ⑩ 587 × 4   ⑪ 184 × 7   ⑫ 214 × 8

⑬ 757 × 8   ⑭ 259 × 7   ⑮ 297 × 8   ⑯ 751 × 9

⑰ 457 × 7   ⑱ 339 × 8   ⑲ 134 × 8   ⑳ 611 × 3

㉑ 578 × 9   ㉒ 247 × 5   ㉓ 188 × 6   ㉔ 968 × 6

㉕ 499 × 9   ㉖ 670 × 4   ㉗ 429 × 3   ㉘ 862 × 5

㉙ 285 × 6   ㉚ 488 × 9   ㉛ 693 × 6   ㉜ 722 × 9

㉝ 457 × 9   ㉞ 767 × 3   ㉟ 312 × 9   ㊱ 691 × 3

BE THERE OR B²: SQUARING 2-DIGIT NUMBERS

Squaring numbers in your head (multiplying a number by itself) is one of the easiest yet most impressive feats of mental calculation you can do. I can still recall where I was when I discovered how to do it. I was 14, sitting on a bus on the way to visit my father at work in downtown Cleveland. It was a trip I made often, so my mind began to wander. I'm not sure why, but I began thinking about the numbers that add up to 20. How large could the product of two such numbers get?

I started in the middle with 10×10 (or 10^2), the product of which is 100. Next, I multiplied $9 \times 11 = 99$, $8 \times 12 = 96$, $7 \times 13 = 91$, $6 \times 14 = 84$, $5 \times 15 = 75$, $4 \times 14 = 56$, and so on. I noticed that the products were getting smaller, and their difference from 100 was 1, 4, 9, 16, 25, 36—or 1^2, 2^2, 3^2, 4^2, 5^2, 6^2 (see Figure 2-1, page 30).

FIGURE 2-1

Numbers that add to 20		Distance from 10	Their Product	Product's difference from 10^2
10	10	0	100	0
9	11	1	99	1
8	12	2	96	4
7	13	3	91	9
6	14	4	84	16
5	15	5	75	25
4	16	6	64	36
3	17	7	51	49
2	18	8	36	64
1	19	9	19	81

I found this pattern astonishing. Next I tried numbers that add to 26 and got similar results. First I worked out $13^2 = 169$, then computed $12 \times 14 = 168$, $11 \times 15 = 165$, $10 \times 16 = 160$, $9 \times 17 = 153$, and so on. Just as before, the distance these products were from 169 was 1^2, 2^2, 3^2, 4^2, and so on (see Figure 2-2, below).

There is actually a simple algebraic explanation for this phenomenon (see the last section of this chapter). At the time, I didn't know my algebra well enough to prove that this pattern would always occur, but I experimented with enough examples to become convinced of it.

Then I realized this pattern could help me square numbers more easily.

FIGURE 2-2

Numbers that add to 26		Distance from 13	Their Product	Product's difference from 13^2
13	13	0	169	0
12	14	1	168	1
11	15	2	165	4
10	16	3	160	9
9	17	4	153	16
8	18	5	146	25

Suppose I wanted to square the number 13, I said to myself. Instead of multiplying 13 × 13, why not get an approximate answer by using two numbers that are easier to multiply but also add up to 26? I chose 10 × 16 = 160. To get an answer, I just added 3^2 (since 10 and 16 are each 3 away from 13). Since $3^2 = 9$, $13^2 = 160 + 9 = 169$. Neat!

This method is diagrammed as follows:

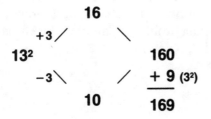

Now let's see how this works for another square:

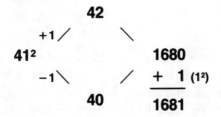

To square 41, subtract 1 to obtain 40 and add 1 to obtain 42. Next multiply 40 × 42. Don't panic! This is simply a 2-by-1 multiplication problem (specifically, 4 × 42) in disguise. Since 4 × 42 = 168, 40 × 42 = 1680. Almost done! All you have to add is the square of 1 (the number by which you went up and down from 41), giving you 1680 + 1 = 1681.

Can squaring a 2-digit number be this easy? Yes, with this method and a little practice, it can. And it works whether you initially round down or round up. For example, let's examine 77^2, working it out both by rounding up and by rounding down:

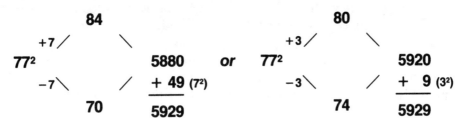

In this instance the advantage of rounding up is that you are virtually done as soon as you have completed the multiplication problem because it is simple to add 9 to a number ending in 0!

In fact, for all 2-digit squares, I always round up or down to the nearest multiple of 10. So if the number to be squared ends in 6, 7, 8, or 9, round up, and if the number to be squared ends in 1, 2, 3, 4, round down. (If the number ends in 5, do both!) With this strategy you will add only the numbers 1, 4, 9, 16, or 25 to your first calculation.

Let's try another problem. Calculate 56^2 in your head before looking at how we did it, below:

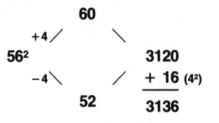

Squaring numbers that end in 5 is even easier. Since you will always round up and down by 5, the numbers to be multiplied will both be multiples of 10. Hence, the multiplication and the addition are especially simple. We have worked out 85^2 and 35^2, below:

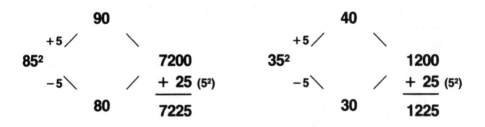

When you are squaring a number that ends in 5, rounding up and down allows you to blurt out the first part of the answer immediately and then finish it with 25. For example, if you want to compute 75^2, rounding up to 80 and down to 70 will give you "Fifty-six hundred . . . and twenty-five!"

For numbers ending in 5, you should have no trouble beating someone with a calculator, and with a little practice with the other squares, it won't be long before you can beat the calculator with *any*

2-digit square number. Even large numbers are not to be feared. You can ask someone to give you a really big 2-digit number, something in the high 90s, and it will sound as though you've chosen an impossible problem to compute. But, in fact, these are even easier because they allow you to round up to 100.

Let's say your audience gives you 96^2. Try it yourself, and then check how we did it:

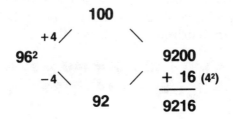

Wasn't that easy? You should have rounded up by 4 to 100 and down by 4 to 92, and then multiplied 100×92 for 9200. At this point you can say out loud, "Ninety-two hundred," and then finish up with "sixteen" and enjoy the applause!

Exercises: 2-Digit Squares

Compute the following:

① 14^2	② 27^2	③ 65^2	④ 89^2	⑤ 98^2
⑥ 31^2	⑦ 41^2	⑧ 59^2	⑨ 26^2	⑩ 53^2
⑪ 21^2	⑫ 64^2	⑬ 42^2	⑭ 55^2	⑮ 75^2
⑯ 45^2	⑰ 84^2	⑱ 67^2	⑲ 103^2	⑳ 208^2

WHY THESE TRICKS WORK

This section is presented for teachers, students, math buffs, and anyone curious as to *why* our methods work. Some people may find the theory as interesting as the application. Fortunately, you need not understand *why* our methods work in order to understand *how* to apply them. All magic tricks have a rational explanation behind them, and mathemagic tricks are no different. It is here that the mathemagician reveals his deepest secrets!

In this chapter on multiplication problems, the *distributive law* is what allows us to break down problems into their component parts. The distributive law states that for any numbers *a*, *b*, and *c*:

$$(b + c) \times a = (b \times a) + (c \times a)$$

That is, the outside term, *a*, is distributed, or separately applied, to each of the inside terms, *b* and *c*. For example, in our first mental multiplication problem of 42 × 7, we arrived at the answer by treating 42 as 40 + 2, then distributing the 7 as follows:

$$42 \times 7 = (40 + 2) \times 7 = (40 \times 7) + (2 \times 7) =$$
$$280 + 14 = 294$$

You may wonder *why* the distributive law works in the first place. To understand it intuitively, imagine having 7 bags, each containing 42 coins, 40 of which are gold and 2 of which are silver. How many coins do you have altogether? There are two ways to arrive at the answer. In the first place, by the very definition of multiplication, there are 42 × 7 coins. On the other hand, there are 40 × 7 gold coins and 2 × 7 silver coins. Hence, we have (40 × 7) + (2 × 7) coins altogether. By answering our question two ways, we have 42 × 7 = (40 × 7) + (2 × 7). Notice that the numbers 7, 40, and 2 could be replaced by any numbers (*a*, *b*, or *c*) and the same logic would apply. That's why the distributive law works!

By using similar logic we can derive other distributive laws, such as:

$$a \times (b - c) = (a \times b) - (a \times c)$$

For instance, since 78 = 80 − 2, 78 × 9 = (80 − 2) × 9, which from the rule above equals (80 × 9) − (2 × 9). To see this intuitively, imagine having 9 bags, each containing 80 coins, 2 of which are made of chocolate. How many non-chocolate coins do you have? Since each bag contains 78 non-chocolate coins, there are 9 × 78 non-chocolate coins. On the other hand, there are 9 × 80 coins altogether but we must eliminate the 9 × 2 chocolate coins (yum!), leaving us with (9 × 80) − (9 × 2) non-chocolate coins.

Using similar reasoning with gold, silver, and platinum coins we can derive:

$$(b + c + d) \times a = (b \times a) + (c \times a) + (d \times a)$$

Hence, to do the problem 326×7, we break up 326 as $300 + 20 + 6$, then distribute the 7, as follows: $326 \times 7 = (300 + 20 + 6) \times 7 = (300 \times 7) + (20 \times 7) + (6 \times 7)$, which we then add up to get our answer.

As for squaring, the following algebra justifies my method: For any numbers A and d

$$A^2 = (A + d) \times (A - d) + d^2$$

Here, A is the number being squared; d can be any number, but I choose it to be the d-istance from A and the nearest multiple of 10. Hence, for $(77)^2$, I set $d = 3$ and our formula tells us, $77^2 = (77 + 3) \times (77 - 3) + 3^2 = (80 \times 74) + 9 = 5929$.

The following algebraic relationship also works to explain my squaring method:

$$(z + d)^2 = z^2 + 2zd + d^2 = z(z + 2d) + d^2$$

Hence, to square 41, we set $z = 40$ and $d = 1$ to get:

$$(41)^2 = (40 + 1)^2 = 40 \times (40 + 2) + 1^2 = 1681$$

Similarly,

$$(z - d)^2 = z(z - 2d) + d^2$$

To find 77^2 when $z = 80$ and $d = 3$,

$$(77)^2 = (80 - 3)^2 = 80 \times (80 - 6) + 3^2 =$$
$$80 \times 74 + 9 = 5929$$

Zerah Colburn: Entertaining Calculations

One of the first lightning calculators to capitalize on his talent was Zerah Colburn (1804–1839), an American farmer's son from Vermont who learned the multiplication tables to 100 before he could even read or write. By the age of six, young Zerah's father took him on the road where his performances generated enough capital to send him to school in Paris and London. By age eight he was internationally famous, performing lightning calculations in England, and was described in the *Annual Register* as "the most singular phaenomenon in the history of the human mind that perhaps ever existed." No less than Michael Faraday and Samuel Morse admired him.

No matter where he went Colburn met all challengers with speed and precision. He tells us in his autobiography of one set of problems he was given in New Hampshire in June 1811: "How many days and hours since the Christian Era commenced, 1811 years? Answered in twenty seconds. 661,015 days, 15,864,360 hours. How many seconds in eleven years? Answered in four seconds; 346,896,000." Colburn used the same techniques described in this book to compute problems given to him entirely in his head. For example, he would factor large numbers into smaller numbers and then multiply: Colburn once multiplied 21,734 × 543 by factoring 543 into 181 × 3. He then multiplied 21,734 × 181 to arrive at 3,933,854, and finally multiplied that figure by 3, for a total of 11,801,562.

As is often the case with lightning calculators, interest in Colburn's amazing skills diminished with time, and by the age of 20 he had returned to America and become a Methodist preacher. He died at a youthful 35. In summarizing his skills as a lightning calculator, and the advantage such an ability affords, Colburn reflected: "True, the method . . . requires a much larger number of figures than the common Rule, but it will be remembered that pen, ink and paper cost Zerah very little when engaged in a sum."

3

New and Improved Products: Intermediate Multiplication

Mathemagics really gets exciting when you perform in front of an audience. I experienced my first taste of public adulation in eighth grade, at the fairly advanced age of 12. Many mathemagicians begin even earlier. Zerah Colburn (1804–1839), for example, reportedly could do lightning calculations before he could read or write, and was entertaining audiences by the age of six! (See the sidebar on Colburn on page 36.) My algebra teacher had just done a problem on the board for which the answer was 108^2. Not content to stop there, I blurted out, "108 squared is simply 11,664!"

The teacher did the calculation on the board and arrived at the same answer. Looking a bit startled, she said, "Yes, that's right. How did you do it?" So I told her: "I went down 8 to 100 and up 8 to 116. I then multiplied 116 × 100, which is 11,600, and just added the square of 8, which is 64, to the total, to get 11,664."

She had never seen that method before. I was thrilled. Thoughts of "Benjamin's Theorem" popped into my head. I actually believed I had discovered something new. When I finally ran across this method later in a Martin Gardner book on recreational math, *Mathematical Carnival* (1965), it ruined my day! Still, the fact that I had developed something on my own gave my confidence a real boost.

You too can impress your friends (or teachers) with some fairly amazing mental multiplication. At the end of the last chapter you learned how to multiply a 2-digit number by itself. In this chapter you will learn how to multiply two different 2-digit numbers, a challenging yet more creative task. You will then try your hand—or, more accurately, your brain—at 3-digit squares. You do not have to know how to do 2-by-2 multiplication problems to tackle 3-digit squares, so you can learn either one first.

2-BY-2 MULTIPLICATION PROBLEMS

When squaring 2-digit numbers, the method is always the same. When multiplying two different 2-digit numbers, however, you can use lots of different methods to arrive at the same answer. For me, this is where the *fun* begins.

The first method you will learn is the *"addition method,"* which can be used to solve all 2-by-2 multiplication problems.

The Addition Method

To use the addition method to multiply any two 2-digit numbers, all you need to do is perform two 2-by-1 multiplication problems and add the results together. For example:

$$
\begin{array}{rr}
 & \mathbf{46} \\
\times & \mathbf{42\ (40 + 2)} \\
\hline
\mathbf{40 \times 46 =}\quad & \mathbf{1840} \\
\mathbf{2 \times 46 = +} & \mathbf{92} \\
\hline
 & \mathbf{1932}
\end{array}
$$

Here you break up 42 into 40 and 2, two numbers that are easy to multiply by. Then you multiply 40×46, which is just 4×46 with a 0 tacked on the end, or 1840. Then you multiply $2 \times 46 = 92$. Finally, you add $1840 + 92 = 1932$, as diagrammed above.

Here's another way to do the same problem:

$$
\begin{array}{rr}
 & \mathbf{46\ (40 + 6)} \\
\times & \mathbf{42} \\
\hline
\mathbf{40 \times 42 =}\quad & \mathbf{1680} \\
\mathbf{6 \times 42 = +} & \mathbf{252} \\
\hline
 & \mathbf{1932}
\end{array}
$$

The catch here is that multiplying 42×6 is harder to do than multiplying 46×2, as in the first problem. Moreover, adding $1680 + 252$ is more difficult than adding $1840 + 92$. So how do you decide which number to break up? I try to choose the number that will produce the easier *addition* problem. In *most* cases—but not all—you will want to break up the number with the smaller last digit because it usually produces a smaller second number for you to add.

Now try your hand at the following problems:

$$\begin{array}{r} 48 \\ \times\quad 73\ (70 + 3) \\ \hline \end{array}$$

$$\begin{array}{rcr} 70 \times 48 & = & 3360 \\ 3 \times 48 & = & +\ 144 \\ \hline & & 3504 \end{array}$$

$$\begin{array}{r} 81\ (80 + 1) \\ \times\quad 59 \\ \hline \end{array}$$

$$\begin{array}{rcr} 80 \times 59 & = & 4720 \\ 1 \times 59 & = & +\ \ 59 \\ \hline & & 4779 \end{array}$$

The last problem illustrates why numbers that end in 1 are especially attractive to break up. If both numbers end in the *same* digit, you should break up the larger number as illustrated below:

$$\begin{array}{r} 84\ (80 + 4) \\ \times\quad 34 \\ \hline \end{array}$$

$$\begin{array}{rcr} 80 \times 34 & = & 2720 \\ 4 \times 34 & = & +\ 136 \\ \hline & & 2856 \end{array}$$

If one number is *much larger* than the other, it often pays to break up the larger number, even if it has a larger last digit. You will see what I mean when you try the following problem two different ways:

$$\begin{array}{r} 74\ (70 + 4) \\ \times\quad 13 \\ \hline \end{array} \qquad\qquad \begin{array}{r} 74 \\ \times\quad 13\ (10 + 3) \\ \hline \end{array}$$

$$\begin{array}{rcr} 70 \times 13 & = & 910 \\ 4 \times 13 & = & +\ 52 \\ \hline & & 962 \end{array} \qquad \begin{array}{rcr} 10 \times 74 & = & 740 \\ 3 \times 74 & = & 222 \\ \hline & & 962 \end{array}$$

Did you find the first method to be faster than the second? I did.

Here's another exception to the rule of breaking up the number with the smallest last digit. When you multiply a number in the 50s by an even number, you'll want to break up the number in the 50s:

$$
\begin{array}{r}
84 \\
\times \quad 59 \ (50 + 9) \\
\hline
50 \times 84 = \quad 4200 \\
9 \times 84 = + \ 756 \\
\hline
4956
\end{array}
$$

The last digit of the number 84 is smaller than the last digit of 59, but if you break up 59, your product will be a multiple of 100, just as 4200 is, in the example above. This makes the subsequent addition problem much easier.

Now try an easy problem of a different sort:

$$
\begin{array}{r}
42 \\
\times \ 11 \ (10 + 1) \\
\hline
10 \times 42 = \quad 420 \\
1 \times 42 = + \ 42 \\
\hline
462
\end{array}
$$

Though the calculation above is pretty simple, there is an even *easier* and *faster* way to multiply any 2-digit number by 11. This is mathemagics at its best; you won't believe your eyes when you see it.

Here's how it works. Suppose you have a 2-digit number whose digits add up to 9 or less. To multiply this number by 11, merely add the two digits together and insert the total between the original two digits. For example, in the problem above, $4 + 2 = 6$. If you place the 6 between the 4 and the 2, it gives you 462, the answer to the problem!

$$
\begin{array}{ccc}
42 & 4 \_ 2 & 462 \\
\times \ 11 & \quad 6 &
\end{array}
$$

Try 54 × 11 by this method.

$$
\begin{array}{ccc}
\mathbf{54} & \mathbf{5\_\_4} & \mathbf{594} \\
\mathbf{\times\ 11} & \mathbf{9} &
\end{array}
$$

What could be simpler? All you had to do was place the 9 between the 5 and the 4 to give you the final answer of 594.

You may wonder what happens when the two numbers add up to a number larger than 9. In such cases, increase the tens digit by 1, then insert the last digit of the sum between the two numbers, as before. For example, when multiplying 76 × 11, 7 + 6 = 13, you increase the 7 in 76 to 8 (86), and then insert the 3 between the 8 and the 6, giving you 836, the final answer.

$$
\begin{array}{ccc}
\mathbf{76} & \mathbf{7\_\_6} & \mathbf{836} \\
\mathbf{\times\ 11} & \mathbf{1\ 3} &
\end{array}
$$

Your turn:

$$
\begin{array}{ccc}
\mathbf{68} & \mathbf{6\_\_8} & \mathbf{748} \\
\mathbf{\times\ 11} & \mathbf{1\ 4} &
\end{array}
$$

Once you get the hang of this trick you will never multiply by 11 any other way. Try a few more problems and then check your answers with ours in the back of the book.

Exercises: Multiplying by 11

①	②	③
35	**48**	**94**
× 11	**× 11**	**× 11**

Piece of cake! But I can't say the same for the next problem, which is a real killer. Try to do it in your head, looking back at the problem if necessary. If you have to start over a couple of times, that's okay.

$$
\begin{array}{rl}
\mathbf{89} & \\
\mathbf{\times\ \ \ 72}\ \mathbf{(70 + 2)} & \\
\hline
\end{array}
$$

$$
\begin{array}{rl}
\mathbf{70 \times 89 =} & \mathbf{6230} \\
\mathbf{2 \times 89 =} & \mathbf{+\ 178} \\
\hline
& \mathbf{6408}
\end{array}
$$

If you got the right answer the first or second time, pat yourself on the back. The 2-by-2 multiplication problems really do not get any tougher than this. If you did not get the answer right away, don't worry. In the next two sections, I'll teach you a couple of much easier strategies for dealing with problems like this. But before you read on, practice the addition method on the following multiplication problems.

Addition-Method Multiplication Problems

①
$$\begin{array}{r} 31 \\ \times\ 41 \\ \hline \end{array}$$

②
$$\begin{array}{r} 27 \\ \times\ 18 \\ \hline \end{array}$$

③
$$\begin{array}{r} 59 \\ \times\ 26 \\ \hline \end{array}$$

④
$$\begin{array}{r} 53 \\ \times\ 58 \\ \hline \end{array}$$

⑤
$$\begin{array}{r} 77 \\ \times\ 43 \\ \hline \end{array}$$

⑥
$$\begin{array}{r} 23 \\ \times\ 84 \\ \hline \end{array}$$

⑦
$$\begin{array}{r} 62 \\ \times\ 94 \\ \hline \end{array}$$

⑧
$$\begin{array}{r} 88 \\ \times\ 76 \\ \hline \end{array}$$

⑨
$$\begin{array}{r} 92 \\ \times\ 35 \\ \hline \end{array}$$

⑩
$$\begin{array}{r} 34 \\ \times\ 11 \\ \hline \end{array}$$

⑪
$$\begin{array}{r} 85 \\ \times\ 11 \\ \hline \end{array}$$

The Subtraction Method

The subtraction method really comes in handy when one of the numbers you want to multiply ends in 8 or 9. The following problem illustrates what I mean:

$$\begin{array}{r} \textbf{59 (60 - 1)} \\ \times\quad \textbf{17} \\ \hline \end{array}$$

$$\begin{array}{rr} \textbf{60} \times \textbf{17} = & \textbf{1020} \\ \textbf{- 1} \times \textbf{17} = - & \textbf{17} \\ \hline & \textbf{1003} \end{array}$$

Now let's do the killer problem from the end of the last section:

$$\begin{array}{r} \textbf{89 (90 - 1)} \\ \times\quad \textbf{72} \\ \hline \end{array}$$

$$\begin{array}{rr} \textbf{90} \times \textbf{72} = & \textbf{6480} \\ \textbf{- 1} \times \textbf{72} = - & \textbf{72} \\ \hline & \textbf{6408} \end{array}$$

Wasn't that a whole lot easier? Now, here's a problem where one number ends in 8:

$$
\begin{array}{r}
\textbf{88 (90 − 2)} \\
\times \quad \textbf{23} \\
\hline
\textbf{90} \times \textbf{23} = \quad \textbf{2070} \\
\textbf{− 2} \times \textbf{23} = \textbf{−} \quad \textbf{46} \\
\hline
\textbf{2024}
\end{array}
$$

In this case you should treat 88 as 90 − 2, then multiply 90 × 23 = 2070. But you multiplied by too much. How much? By 2 × 23, or 46. So subtract 46 from 2070 to arrive at 2024, the final answer.

I want to emphasize here that it is important to work these problems out in your head and not simply look to see how we did it in the diagram. Go through them and say the steps to yourself or even out loud to reinforce the mental picture and process.

Not only do I use the subtraction method with numbers that end in 8 or 9, but also for numbers in the high 90s because 100 is such a convenient number to multiply. For example, if someone asked me to multiply 96 × 73, I would immediately round 96 up to 100:

$$
\begin{array}{r}
\textbf{96 (100 − 4)} \\
\times \quad \textbf{73} \\
\hline
\textbf{100} \times \textbf{73} = \quad \textbf{7300} \\
\textbf{− 4} \times \textbf{73} = \textbf{−} \quad \textbf{292} \\
\hline
\textbf{7008}
\end{array}
$$

When the subtraction component of a multiplication problem requires you to borrow a number, using complements can help you arrive at the answer more quickly. You'll see what I mean as you work your way through the problems below. For example, subtract 340 − 78. The difference between 40 and 78 is 38. The complement of 38 is 62. So 62 will be the final two digits of the answer, which is 262:

$$
\begin{array}{rl}
\textbf{340} & \textbf{78 − 40 = 38} \\
\textbf{− 78} & \textbf{Complement of 38 = 62} \\
\hline
\textbf{262} &
\end{array}
$$

$$88 \ (90 - 2)$$
$$\times \quad 76$$

$$90 \times 76 = \quad 6840$$
$$-2 \times 76 = \ - \ 152$$

There are two ways to perform the subtraction component of this problem. Here's the long way:

$$6840 \quad -200 \qquad 6640 \quad +48 \qquad 6688$$
$$- \ 152 \quad \longrightarrow \quad + \quad 48 \quad \longrightarrow$$

The *short* way is to subtract $52 - 40 = 12$ and then find the complement of 12, which is 88. Since 52 is greater than 40, you know the answer will be 66 hundred and something. You can then plug in 88 as the last two digits of your answer.

Try this one:

$$67$$
$$\times \quad 59 \ (60 - 1)$$

$$60 \times 67 = \quad 4020$$
$$-1 \times 67 = \ - \quad 67$$
$$\overline{ 3953}$$

Again, you can see that the answer will be 3900 and something. Because $67 - 20 = 47$, the complement 53 means the answer is 3953.

As you may have realized, this method works with any subtraction problem that requires you to borrow a number, not just those that are part of a multiplication problem. Here is further proof, if you need it, that complements are a very powerful tool in mathemagics. Master this technique and pretty soon people will be complimenting you!

Subtraction-Method Multiplication Problems

①	②	③	④
29	98	47	68
× 45	× 43	× 59	× 38

⑤
$$96 \times 29$$

⑥
$$79 \times 54$$

⑦
$$37 \times 19$$

⑧
$$87 \times 22$$

⑨
$$85 \times 38$$

⑩
$$57 \times 39$$

⑪
$$88 \times 49$$

The Factoring Method

The factoring method, my favorite method of multiplying 2-digit numbers, involves no addition or subtraction at all. You use it when one of the numbers in a 2-digit multiplication problem can be factored into small numbers.

To factor a number means to break it down into smaller numbers which, when multiplied together, give the original number. For example, the number 24 can be factored into 8×3 or 6×4. (It can also be factored into 12×2, but single-digit factors are easier to work with, and your goal is to make life easier for yourself.)

Here are some other examples of factored numbers:

$$42 = 7 \times 6$$
$$63 = 9 \times 7$$
$$84 = 7 \times 6 \times 2 \text{ or } 7 \times 4 \times 3$$

To see how factoring makes multiplication easier, consider the following problem, repeated from the section on the addition method earlier in this chapter:

$$46 \times 42$$

Previously we solved this problem by multiplying 46×40 and 46×2 and adding the products together. To use the factoring method, treat 42 as 7×6 and begin by multiplying 46×7, which is 322. Then multiply $322 \times 6 = 1932$ for the final answer. You already know how to do 2-by-1 and 3-by-1 multiplication problems, so this should be a cinch:

$$46 \times 42 = 46 \times (7 \times 6) = (46 \times 7) \times 6 =$$
$$322 \times 6 = 1932$$

Of course, this problem could also have been solved by reversing the factors of 42:

$$46 \times 42 = 46 \times (6 \times 7) = (46 \times 6) \times 7 =$$
$$276 \times 7 = 1932$$

But as you can see, it is easier to multiply 322×6 than it is to multiply 276×7. In most cases, I like to use the larger factor in solving the initial 2-by-1 problem and to reserve the smaller factor for the 3-by-1 component of the problem.

Factoring allows you to do what you need to do: simplify from larger to smaller numbers, and from harder to easier problems. The advantage of the factoring method in mental calculation is you do not have to hold much in memory. Let's look at another problem by way of example:

$$75 \times 63 = 75 \times (9 \times 7) = (75 \times 9) \times 7 =$$
$$675 \times 7 = 4725$$

As before, you simplify this 2-by-2 problem by factoring 63 into 9×7 and then multiplying 75 by these factors.

You may have noticed, however, that there is another way to factor this problem. You can also treat 75 as 25×3, or $5 \times 5 \times 3$:

$$63 \times 75 = 63 \times (5 \times 5 \times 3) = (63 \times 5) \times 5 \times 3 =$$
$$315 \times 5 \times 3 = 1575 \times 3 = 4725$$

This particular problem works out nicely because the multiplication problems are easy to compute. But given the choice, I still prefer working with two factors instead of three because there is less to remember.

Try the following problem for practice:

$$57 \times 24 = 57 \times 8 \times 3 = 456 \times 3 = 1368$$

You could have factored 24 as 6×4 for another easy computation:

$$57 \times 24 = 57 \times 6 \times 4 = 342 \times 4 = 1368$$

By way of comparison, notice how much easier this method is than the addition method:

$$
\begin{array}{r}
\mathbf{57} \\
\times \quad \mathbf{24}\ \mathbf{(20 \times 4)} \\
\hline
\mathbf{20 \times 57 =} \quad \mathbf{1140} \\
\mathbf{4 \times 57 =} +\ \mathbf{228} \\
\hline
\mathbf{1368}
\end{array}
$$

Here you have to perform two 2-by-1 problems and then add. With the factoring method, you have just two multiplication problems: a 2-by-1 and a 3-by-1. You use less memory with the factoring method.

Remember that "killer" problem at the end of the addition section? Here it is again:

$$
\begin{array}{r}
\mathbf{89} \\
\times\ \mathbf{72} \\
\hline
\end{array}
$$

We tackled that problem easily enough with the subtraction method, but factoring works even faster:

$$\mathbf{89 \times 72 = 89 \times 9 \times 8 = 801 \times 8 = 6408}$$

The problem is especially easy because of the 0 in the middle of 801. Our next example illustrates that it sometimes pays to factor the numbers in an order that exploits this situation. Let's look at two ways of computing 67×42:

$$\mathbf{67 \times 42 = 67 \times 7 \times 6 = 469 \times 6 = 2814}$$

$$\mathbf{67 \times 42 = 67 \times 6 \times 7 = 402 \times 7 = 2814}$$

Ordinarily you would factor 42 into 7×6, as in the first example, so that you could multiply the larger factor first. But the problem is easier to solve if you factor 42 into 6×7 because it creates a number with a 0 in the center, which is easier to multiply. I call such numbers "friendly products."

Look for the friendly product in the problem done two ways below:

$$43 \times 56 = 43 \times 8 \times 7 = 344 \times 7 = 2408$$

$$43 \times 56 = 43 \times 7 \times 8 = 301 \times 8 = 2408$$

Isn't the second way easier?

The trick to using the factoring method is to equip yourself to find friendly products whenever you can. The following list should help. I don't expect you to memorize it so much as to familiarize yourself with it. With practice you will be able to nose out friendly products more often, and the list will become more meaningful.

Numbers with Friendly Products

12:	$12 \times 9 = 108$
13:	$13 \times 8 = 104$
15:	$15 \times 7 = 105$
17:	$17 \times 6 = 102$
18:	$18 \times 6 = 108$
21:	$21 \times 5 = 105$
23:	$23 \times 9 = 207$
25:	$25 \times 4 = 100, 25 \times 8 = 200$
26:	$26 \times 4 = 104, 26 \times 8 = 208$
27:	$27 \times 4 = 108$
29:	$29 \times 7 = 203$
34:	$34 \times 3 = 102, 34 \times 6 = 204, 34 \times 9 = 306$
35:	$35 \times 3 = 105$
36:	$36 \times 3 = 108$
38:	$38 \times 8 = 304$
41:	$41 \times 5 = 205$
43:	$43 \times 7 = 301$
44:	$44 \times 7 = 308$
45:	$45 \times 9 = 405$
51:	$51 \times 2 = 102, 51 \times 4 = 204, 51 \times 6 = 306,$
	$51 \times 8 = 408$
52:	$52 \times 2 = 104, 52 \times 4 = 208$
53:	$53 \times 2 = 106$
54:	$54 \times 2 = 108$
56:	$56 \times 9 = 504$

61:     $61 \times 5 = 305$
63:     $63 \times 8 = 504$
67:     $67 \times 3 = 201, 67 \times 6 = 402, 67 \times 9 = 603$
68:     $68 \times 3 = 204, 68 \times 6 = 408$
72:     $72 \times 7 = 504$
76:     $76 \times 4 = 304, 76 \times 8 = 608$
77:     $77 \times 4 = 308$
78:     $78 \times 9 = 702$
81:     $81 \times 5 = 405$
84:     $84 \times 6 = 504$
88:     $88 \times 8 = 704$
89:     $89 \times 9 = 801$

Previously in this chapter you learned how easy it is to multiply numbers by 11. It usually pays to use the factoring method when one of the numbers is a multiple of 11:

$$52 \times 33 = 52 \times 11 \times 3 = 572 \times 3 = 1716$$

$$83 \times 66 = 83 \times 11 \times 6 = 913 \times 6 = 5478$$

Factoring-Method Multiplication Problems

①
$$27 \\ \times 14$$

②
$$86 \\ \times 28$$

③
$$57 \\ \times 14$$

④
$$81 \\ \times 48$$

⑤
$$56 \\ \times 29$$

⑥
$$83 \\ \times 18$$

⑦
$$72 \\ \times 17$$

⑧
$$85 \\ \times 42$$

⑨
$$33 \\ \times 16$$

⑩
$$62 \\ \times 77$$

⑪
$$45 \\ \times 36$$

⑫
$$48 \\ \times 37$$

APPROACHING MULTIPLICATION CREATIVELY

I mentioned at the beginning of the chapter that multiplication problems are fun because they can be solved any number of ways.

Now that you know what I mean, let's apply all three methods explained in this chapter to a single problem, 73 × 49. We'll begin by using the addition method:

$$
\begin{array}{r}
\mathbf{73\ (70 + 3)} \\
\times\quad \mathbf{49} \\
\hline
70 \times 49 = \quad \mathbf{3430} \\
3 \times 49 = +\ \ \mathbf{147} \\
\hline
\mathbf{3577}
\end{array}
$$

Now try the subtraction method:

$$
\begin{array}{r}
\mathbf{73} \\
\times\quad \mathbf{49\ (50 - 1)} \\
\hline
50 \times 73 = \quad \mathbf{3650} \\
-\ 1 \times 73 = -\quad \mathbf{73} \\
\hline
\mathbf{3577}
\end{array}
$$

Note that the last two digits of the subtraction could be obtained by adding 50 + (complement of 73) = 50 + 27 = 77 or by simply taking the (complement of 73 − 50) = complement of 23 = 77.

Finally, try the factoring method:

$$\mathbf{73 \times 49 = 73 \times 7 \times 7 = 511 \times 7 = 3577}$$

Congratulations! You have mastered 2-by-2 multiplication and now have all the basic skills you need to be a fast mental calculator. All you need to become a *lightning* calculator is *more practice.*

Exercises for 2-by-2 General Multiplication: Anything Goes!

Many of the following exercises can be solved by more than one method. Try computing them in as many ways as you can think of; then check your answers and computations with ours in the back of the book. Our answers suggest various ways the problem can be mathemagically solved, starting with the easiest method.

①	②	③	④	⑤
53	81	73	89	77
× 39	× 57	× 18	× 55	× 36

⑥ 92 × 53   ⑦ 87 × 87   ⑧ 67 × 58   ⑨ 56 × 37   ⑩ 59 × 21

The following 2-by-2s occur as subproblems to problems appearing later in the text when we do 3-by-2s, 3-by-3s, and 5-by-5s. You will see these problems embedded in these larger problems, so you can do them now for practice, or refer back to them when they are used in the larger problems.

⑪ 37 × 72   ⑫ 57 × 73   ⑬ 38 × 63   ⑭ 43 × 76   ⑮ 43 × 75

⑯ 74 × 62   ⑰ 61 × 37   ⑱ 36 × 41   ⑲ 54 × 53   ⑳ 53 × 53

㉑ 83 × 58   ㉒ 91 × 46   ㉓ 52 × 47   ㉔ 29 × 26   ㉕ 41 × 15

㉖ 65 × 19   ㉗ 34 × 27   ㉘ 69 × 78   ㉙ 95 × 81   ㉚ 65 × 47

㉛ 65 × 69   ㉜ 95 × 26   ㉝ 41 × 93

3-DIGIT SQUARES

Squaring 3-digit numbers is an impressive feat of mental prestidigitation. Just as you square 2-digit numbers by rounding up or down to the nearest multiple of 10, to square 3-digit numbers, you round up or down to the nearest multiple of 100. Take 193^2:

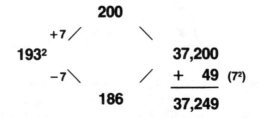

By rounding up to 200 and down to 186, you've transformed a 3-by-3 multiplication problem into a far simpler 3-by-1 problem. After all, 200 × 186 is just 2 × 186 = 372 with two zeros attached. Almost done! Now all you have to add is 7^2 = 49 to arrive at 37,249.

Now try squaring 706:

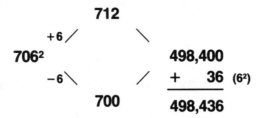

Rounding *down* by 6 to 700 requires you to round *up* by 6 to 712. Since 712 × 7 = 4984 (a simple 3-by-1 problem), 712 × 700 = 498,400. After adding 6^2 = 36 you arrive at 498,436.

These last problems are not terribly hard because they don't require you to carry any numbers. Moreover, you know the answers to 6^2 and 7^2 by heart. Squaring a number that's farther away from a multiple of 100 is a tougher proposition. Try your hand at 314^2:

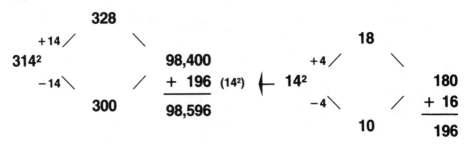

For this 3-digit square, go down 14 to 300 and up 14 to 328, then multiply 328 × 3 = 984. Tack on two 0s to arrive at 98,400. Then add the square of 14. If 14^2 = 196 comes to you in a flash (through memory or calculation), you're in good shape. Just add 98,400 + 196 to arrive at 98,596. If you need time to compute 14^2, repeat the number 98,400 to yourself a few times before you go on. Otherwise you might compute $14^2 = 196$ and forget what number to add to it.

The farther away you get from a multiple of 100, the more difficult squaring a 3-digit number becomes. Try 529^2:

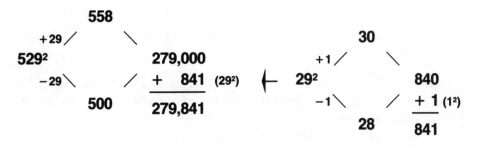

If you have an audience you want to impress, you can say 279,000 out loud before you compute 29^2. But this will not work for every problem. For instance, try squaring 636:

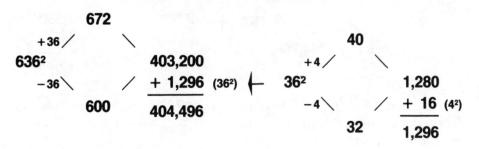

Now your brain is really working, right? The key here is to repeat 403,200 to yourself several times. Then square 36 to get 1296 in the usual way. The hard part comes in adding 1296 to 403,200. Do it one digit at a time, left to right, to arrive at your answer of 404,496. Take our word that as you become more familiar with 2-digit squares, these 3-digit problems get easier.

Here's an even tougher problem, 863^2:

The first problem is deciding what numbers to multiply together. Clearly one of the numbers will be 900, and the other number will be in the 800s. But what are the last two digits? You can compute them two ways:

1. **The hard way:** The difference between 863 and 900 is 37 (the complement of 63). Subtract 37 from 863 to arrive at 826.
2. **The easy way:** Double the number 63 to get 126, and take the last two digits to give you 826.

Here's why the easy way works. Because both numbers are the same distance from 863, their sum must be twice 863, or 1726. One of your numbers is 900, so the other must be 826.

You then compute the problem like this:

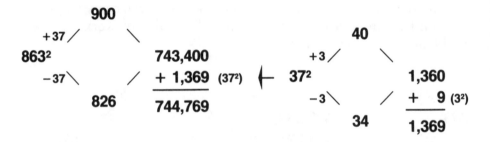

If you find it impossible to remember 743,400 after squaring 37, fear not. In Chapter 7, you will learn a memory system that will make remembering such numbers a breeze.

Try your hand at squaring 359, the hardest problem yet:

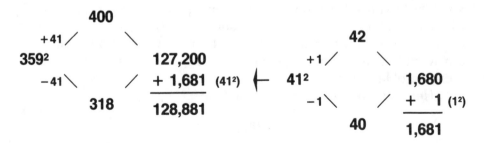

To obtain 318, either subtract 41 (the complement of 59) from 359, or multiply $2 \times 59 = 118$ and use the last two digits. Next multiply $400 \times 318 = 127,200$. Adding 41^2, or 1681, gives you 128,881.

Whew! They don't get much harder than that! If you got it right the first time, take a bow!

Let's finish with an easy one 987^2:

$$987^2 \quad \begin{matrix} & 1000 \\ +13 \nearrow & & \searrow \\ & & 974{,}000 \\ -13 \searrow & & \nearrow \quad + \quad 169 \ (13^2) \\ & 974 & 974{,}169 \end{matrix} \longleftarrow 13^2 \quad \begin{matrix} & 16 \\ +3 \nearrow & & \searrow \\ & & 160 \\ -3 \searrow & & \nearrow \quad + \quad 9 \ (3^2) \\ & 10 & 169 \end{matrix}$$

Exercises for 3-Digit Squares

① 409^2 ② 805^2 ③ 217^2 ④ 896^2

⑤ 345^2 ⑥ 346^2 ⑦ 276^2 ⑧ 682^2

⑨ 431^2 ⑩ 781^2 ⑪ 975^2

What's Behind Door Number 1?

The mathematical chestnut of 1991 that got everyone hopping mad was an article in *Parade* magazine by Marilyn vos Savant, the woman listed by the *Guinness Book of World Records* as having the world's highest I.Q. The paradox has come to be known as the Monty Hall problem, and it goes like this.

You are a contestant on *Let's Make a Deal.* Monty Hall allows you to pick one of three doors, behind one of which is the big prize, behind the other two are goats. You pick Door Number 2. But before Monty reveals the prize of your choice, he shows you what you didn't pick behind Door Number 3. It's a goat. Now, in his tantalizing way, Monty gives you another choice. Do you want to stick with Door Number 2, or do you want to risk a chance to see what's behind Door Number 1? What should you do? Assuming that Monty is only going to reveal where the big prize is not, he will always open one of the consolation doors. This leaves two doors, one with the big prize and the other with another consolation. The odds are now 50–50 for your choice, right?

Wrong! The odds that you chose correctly the first time remain 1 in 3. The probability that the big prize is under the other door increases to 2 in 3 because probability must add to 1. Thus, by switching doors you double the odds of winning! (The problem assumes that Monty will always give a player the option to switch, that he will always reveal a nonwinning door, and that when your first pick is correct he will choose a nonwinning door at random.) Think of playing the game with 10 doors and after your pick he reveals 8 other nonwinning doors. You would, of course, switch. People confuse this problem for a variant: If Monty Hall does not know where the grand prize is, and reveals Door 3 which happens to contain a goat (though it might have contained the prize), then Door 1 has a 50% chance of being correct. This result is so counterintuitive that Marilyn Savant received piles of letters, many from scientists and even mathematicians, telling her she didn't deserve her honorific title of highest I.Q. They were all wrong.

4

Divide and Conquer: Mental Division

Mental division is a particularly handy skill to have, both in business and in daily life. How many times a week are you confronted with situations that call on you to divide things up evenly, such as a check at a restaurant? This same skill comes in handy when you want to figure out the cost per unit of a case of dog food on sale, or to split the pot in poker, or to figure out how many gallons of gas you can buy with a $20 bill. The ability to divide in your head can save you the inconvenience of having to pull out a calculator every time you need to compute something.

With mental division, the left-to-right method of calculation comes into its own. This is the same method we all learned in school, so you will be doing what comes naturally. I remember as a kid thinking that this left-to-right method of division is the way all arithmetic should be done. I have often speculated that if the schools could have figured out a way to teach division right-to-left, they probably would have done so!

1-DIGIT DIVISION

The first step when dividing mentally is to figure out how many digits will be in your answer. To see what I mean, try the following problem on for size:

179 ÷ 7

To solve $179 \div 7$, we're looking for a number, Q, such that 7 times Q is 179. Now, since 179 lies between $7 \times 10 = 70$, and $7 \times 100 =$

700, Q must lie between 10 and 100, which means our answer must be a 2-digit number. Knowing that, divide 7 into the first two digits of 179, namely 17, to arrive at your first digit. You know that $7 \times 2 = 14$ and $7 \times 3 = 21$, therefore your first digit is 2. Just as you would using pencil and paper, you then multiply $7 \times 2 = 14$, subtract $17 - 14 = 3$, bring down the 9 from 179 and then divide 7 into 39. Since $7 \times 5 = 35$, the answer is 25 with a remainder of 4.

As you compute this problem mentally, you can keep the 20 in mind by holding it on your fingers—just put out the first two fingers on your left hand to represent 20. Or you can say "twenty" out loud from the start since the first digit will not change. Then complete the calculation. Here's what the process looks like:

$$
\begin{array}{r}
25 \\
7{\overline{)179}} \\
14 \\
\hline
39 \\
35 \\
\hline
\end{array}
$$

4 — remainder

Answer: 25 with a remainder of 4, or $25\frac{4}{7}$

Let's try a similar division problem using the same method of mental computation:

$$675 \div 8$$

As before, since 675 falls between $8 \times 10 = 80$ and $8 \times 100 = 800$, your answer must be below 100 and therefore is a 2-digit number. To divide 8 into 67, remember that $8 \times 8 = 64$ and $8 \times 9 = 72$. Therefore, your answer is 80 something, a 2-digit number.

But what is that "something"? To find out, subtract 64 from 67 for a remainder of 3. Bring down the 5, then divide 8 into 35. Since $8 \times 4 = 32$, subtract 32 from 35 giving a remainder of 3. The final answer is 84 with a remainder of 3. We can illustrate this problem as follows:

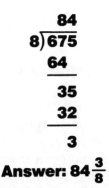

Answer: 84$\frac{3}{8}$

In this problem, as before, you can say the answer out loud as you go along because the first digit will not change. Or you can "hold" the number on your fingers to remember it and say the entire answer at once.

Now let's try a division problem that results in a 3-digit answer:

947 ÷ 4

You can tell right away that your answer will have three digits because 947 falls between 4 × 100 = 400 and 4 × 1000 = 4000. Begin by dividing 4 into 9, which gives you 2 with a remainder of 1. (Raise two fingers on your left hand to help you remember.) Bring down the 4 and divide 4 into 14, giving you 3 with a remainder of 2. (Raise three fingers on your right hand.) Bring down the 7 and divide 4 into 27, to arrive at 6 with a remainder of 3. Holding two numbers on your fingers and two in your head, put them all together to arrive at the answer: 236 with a remainder of 3.

$$
\begin{array}{r}
236 \\
4\overline{)947} \\
8 \\
\hline
14 \\
12 \\
\hline
27 \\
24 \\
\hline
3
\end{array}
$$

Answer: 236$\frac{3}{4}$

Division problems don't get too tough until you divide a 1-digit number into a 4-digit number:

2196 ÷ 5

You can see the answer will have three digits because 2196 is between 5 × 100 = 500 and 5 × 1000 = 5000. Dividing 5 into 21 gives you 4 with a remainder of 1. At this point you know the answer will be "400 something" and you can say "four hundred" out loud or hold the 4 on the fingers of your left hand to help you remember.

Now your problem has been reduced to 400 plus the solution of 196 ÷ 5. (It's helpful to think of the problem this way so you don't forget 6, the ones digit.) Next bring down the 9 and divide 5 into 19. Your answer is 3 (5 × 3 = 15), giving you your tens digit, 3. You can say "four hundred and thirty" or hold the 3 on your right hand. Almost done.

Now subtract 15 from 19, giving you 4, and bring down the 6. Finally, calculate that 5 goes into 46 nine times with a remainder of 1. Now you can say "439 with a remainder of 1, or $439\frac{1}{5}$."

$$
\begin{array}{r}
439 \\
5\overline{)2196} \\
20 \\
\hline
19 \\
15 \\
\hline
46 \\
45 \\
\hline
1
\end{array}
$$

Answer: $439\frac{1}{5}$

THE RULE OF THUMB

When dividing in your head instead of on paper, you may find it difficult to remember parts of the answer as you continue calculating. One option, as you've seen, is to say the answer out loud as you go. But for greater dramatic effect, you may prefer, as I do, to hold the answer on your fingers and say it all together at the end. In that case, you may run into problems remembering digits greater than 5, if, like most of us, you have only 5 fingers on each hand. The solution is to use a special technique borrowed from sign language to remember such numbers. This technique, which I call the "rule of thumb," is most effective for remembering 3-digit and above answers. This technique is not only useful in this chapter but will also come in handy (pardon the pun) in subsequent chapters dealing with larger problems and longer numbers to remember.

You already know that to represent numbers 0 through 5, all you have to do is raise the equivalent number of fingers on your hand. If you get your thumb in on the act, it's just as easy to represent numbers 6 through 9. To remember the number 6, simply put your thumb and little finger together. To remember 7, put your thumb and ring finger together, and so on as illustrated below.

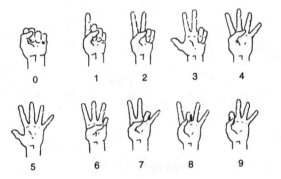

With 3-digit answers, hold the hundreds digit on your left hand and the tens digit on your right. When you get to the ones digit, you've reached the end of the problem (except for a possible remainder). Now pipe up with the number on your left hand, the number on your right hand, the one-digit you've just computed, and the remainder. Presto—you've said your answer!

For practice, try computing the following 4-digit division problem:

$$4579 \div 6$$

$$
\begin{array}{r}
763 \\
6 \overline{)4579} \\
42 \\
\hline
37 \\
36 \\
\hline
19 \\
18 \\
\hline
1
\end{array}
$$

Answer: $763\frac{1}{6}$

In using the rule of thumb to remember the answer, you'll hold the 7 on your left hand by placing your thumb and ring finger together and the 6 on your right hand by placing your thumb and little finger together. Once you've calculated the ones digit (which is 3) and the remainder (which is 1), you can "read" the final answer off your hands from left to right: "7 . . . 6 . . . 3 with a remainder of 1, or $\frac{1}{6}$th."

Some 4-digit division problems yield 4-digit answers. In that case, if you are one of those unfortunate people who have only two hands, you will have to say the thousands digit of the answer out loud and then use the rule of thumb to remember the rest of the answer. For example:

$$8352 \div 3$$

$$
\begin{array}{r}
2784 \\
3\overline{)8352} \\
6 \\
\hline
23 \\
21 \\
\hline
25 \\
24 \\
\hline
12 \\
12 \\
\hline
0
\end{array}
$$

Answer: 2784

For this problem you divide 3 into 8 to get your thousands digit of 2, say "two thousand" out loud, then divide 3 into 2352 in the usual way.

2-DIGIT DIVISION

The rule of thumb is also useful for problems where you must divide a 2-digit number into 3- and 4-digit numbers. The method is the same, although the answer to such problems is a little trickier to compute because the products are larger and the subtraction problems are more complicated. This makes it especially important that you use the rule of thumb so that you don't have to think about the answer as you compute.

It's inevitable that division problems become harder as the number you divide by gets larger. Fortunately, I have some magic up my sleeve to make your life easier.

Let's start with a relatively easy problem first:

597 ÷ 14

To determine the answer, your first step is to ask how many times 14 goes into 59. Because 14 × 4 = 56 you know that the answer is 40 something and so you can say "forty" out loud or just hold the 4 on your left hand.

Next subtract $59 - 56 = 3$, which reduces your problem to dividing 14 into 37. Since $14 \times 2 = 28$, your answer is 42. Subtracting 28 from 37 leaves you a remainder of 9. The process of deriving the solution to this problem may be illustrated as follows:

$$
\begin{array}{r}
42 \\
14\overline{)597} \\
56 \\
\hline
37 \\
28 \\
\hline
9
\end{array}
$$

Answer: $42\frac{9}{14}$

The following problem is slightly harder because the 2-digit divisor (the official term for the number by which you divide) is larger:

$682 \div 23$

You know right away that the answer is a 2-digit number because 682 falls between $23 \times 10 = 230$ and $23 \times 100 = 2300$. To figure out the tens digit of the 2-digit answer you need to ask how many times 23 goes into 68. If you try 3, you'll quickly see that it's slightly too much, as $3 \times 23 = 69$. Now you know that the answer is 20 something and you can say so out loud or hold the 2 on your left hand. Then multiply $23 \times 2 = 46$ and subtract $68 - 46 = 22$. After bringing down the 2, how many times does 23 go into 222? You can see that 10 is close, but slightly too much at 230, so the answer is 9, as $23 \times 9 = 207$. Put the 9 on your right hand (using the rule of thumb), then subtract $222 - 207 = 15$ for your remainder. Read your hands from left to right: 29 and $\frac{15}{23}$.

$$
\begin{array}{r}
29 \\
23\overline{)682} \\
46 \\
\hline
222 \\
207 \\
\hline
15
\end{array}
$$

Answer: $29\frac{15}{23}$

Now consider:

491 ÷ 62

Since 491 falls between $1 \times 62 = 62$ and $10 \times 62 = 620$, your answer will simply be a 1-digit number with a remainder. It makes sense to lop off the last digit of each number to help you "guess-timate" an answer. As it happens, $49 \div 6 = 8\frac{1}{6}$, but when you multiply 62×8, you arrive at 496, which is a little high. Since $62 \times 7 = 434$, the answer is 7 with a remainder of $491 - 434 = 57$, or $7\frac{57}{62}$.

$$
\begin{array}{r}
7 \\
62\overline{)491} \\
434 \\
\hline
57
\end{array}
$$

Answer: $7\frac{57}{62}$

Here's a nifty trick to make problems like this easier. Remember how you first tried multiplying 62×8 but found it came out a little high at 496? Well, that wasn't a wasted effort. Aside from knowing that the answer is 7, you can compute the remainder right away. Since 496 is 5 more than 491, the remainder will be 5 less than 62, the divisor. Since $62 - 5 = 57$, your answer is $7\frac{57}{62}$. The reason this trick works is because $491 = (62 \times 8) - 5 = 62 \times (7 + 1) - 5 = 62 \times 7 + 62 - 5 = 62 \times 7 + 57$.

Your next challenge is to divide a 2-digit number into a 4-digit number:

3657 ÷ 54

Since $54 \times 100 = 5400$, you know your answer will be a 2-digit number. To arrive at the first digit of the answer you need to figure how many times 54 goes into 365 or, as a rough estimate, how many times 5 goes into 36. Knowing that $5 \times 7 = 35$, it pays to try $54 \times 7 = 378$. Once you see that this is too high, you know the answer must be 60 something, which you can say out loud or hold on your hand.

Next, multiply $54 \times 6 = 324$ and subtract $365 - 324 = 41$. (Alternatively, if we noted that 378 was 13 higher than 365, we could have used the trick we used in the last problem for a quotient of 6 with a remainder of 54 − 13 = 41.) Once you bring down the 7, you need to figure out how many times 54 goes into 417. Well, you know there are eight 50s in 400, but you're dividing by 54, so you might try 7. Fortunately, 54 × 7 = 378, which means 7 is the second digit of the answer. Put 7 on your right hand. (Aren't you glad after all these calculations that you put that 6 on your left hand so you didn't have to think about it?) Then subtract 417 − 378 = 39, the remainder.

$$
\begin{array}{r}
67 \\
54{\overline{\smash{)}3657}} \\
3240 \\
\hline
417 \\
378 \\
\hline
39 \\
\end{array}
$$

Answer: $67\frac{39}{54}$

Now try your hand (or hands) at a problem with a 3-digit answer that requires greater use of the rule of thumb:

9467 ÷ 13

$$
\begin{array}{r}
728 \\
13{\overline{\smash{)}9467}} \\
9100 \\
\hline
367 \\
260 \\
\hline
107 \\
104 \\
\hline
3 \\
\end{array}
$$

Answer: $728\frac{3}{13}$

At the conclusion of the problem, you should have a 7 on your left hand, a 2 on your right hand, an 8 in your left brain, and a remainder of 3 in your right brain!

SIMPLIFYING DIVISION PROBLEMS

If by this point you're suffering from brain strain, relax. As promised, I want to share with you a couple of tricks for simplifying certain mental division problems. These tricks are based on the principle of dividing both parts of the problem by a common factor. If both numbers in the problem are even numbers, you can make the problem twice as easy by dividing each number by two before you begin. For example, 858 ÷ 16 has two even numbers, and dividing each by 2 yields the much simpler problem of 429 ÷ 8:

```
        53                              53
   16)858            ⟶            8)429
       80              ÷ 2            40
       ──                            ──
       58                            29
       48                            24
       ──                            ──
       10                             5
```

Answer: 53 10/16   **Answer: 53 5/8**

As you can see, the remainders 10 and 5 are not the same, but if you write the remainder in the form of a fraction you will see that $\frac{10}{16}$ is the same as $\frac{5}{8}$. Therefore, when using this method you must always express the answer in fractional form.

We've done both sets of calculations for you to see how much easier it is. Now you try one for practice:

3618 ÷ 54

```
        67                              67
   54)3618          ⟶           27)1809
      324             ÷ 2            162
      ───                           ───
      378                           189
      378                           189
      ───                           ───
        0                             0
```

Answer: 67

The problem on the right is much easier to calculate mentally. If you're *really* alert, you could divide both sides by 9 to arrive at an even simpler problem: $201 \div 3 = 67$.

Watch for problems that can be divided by 2, *twice*, like $1652 \div 36$:

$$1652 \div 36 \xrightarrow[\div 2]{} 826 \div 18 \xrightarrow[\div 2]{} 413 \div 9 = 9\overline{)413}$$

$$
\begin{array}{r}
45 \\
9\overline{)413} \\
36 \\
\hline
53 \\
45 \\
\hline
8
\end{array}
$$

Answer: $45\dfrac{8}{9}$

Of course, when both numbers end in 0, you can divide each by 10:

$$580 \div 70 \xrightarrow[\div 10]{} 58 \div 7 = 7\overline{)58}$$

$$
\begin{array}{r}
8 \\
7\overline{)58} \\
56 \\
\hline
2
\end{array}
$$

Answer: $8\dfrac{2}{7}$

But if both numbers end in 5, double them and then divide both by 10 to simplify the problem. For example:

$$475 \div 35 \xrightarrow[\times 2]{} 950 \div 70 \xrightarrow[\div 10]{} 95 \div 7 = 7\overline{)95}$$

$$
\begin{array}{r}
13 \\
7\overline{)95} \\
7 \\
\hline
25 \\
21 \\
\hline
4
\end{array}
$$

Answer: $13\dfrac{4}{7}$

Finally, if the divisor ends in 5 and the number you're dividing into ends in 0, multiply both by 2 and then divide by 10, just as you did above:

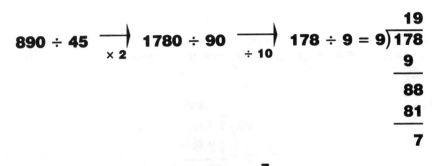

$$\text{Answer: } 19\frac{7}{9}$$

MATCHING WITS WITH A CALCULATOR: LEARNING DECIMALIZATION

When I perform in front of an audience, I invite members of the audience to check my calculations on a calculator. (I confess that I'm so practiced a mathemagician that I usually beat them to the answer.) When you're competing against a calculator, a remainder expressed in fractional form won't do because it doesn't match what the calculator spits out. The solution is to convert fractional remainders into decimal ones. Simply divide the bottom of the fraction, or the denominator, into the top of the fraction, or the numerator. Since the denominator is always larger than the numerator, you will need to append 0s to the numerator after the decimal point.

For example, let's say your answer is $67\frac{11}{17}$ and you want to convert the $\frac{11}{17}$ to a decimal. You'll want to say "67 point . . ." to designate the first part of the answer and the decimal point and give the rest of the answer digit by digit as you go along. In this case you will divide 17 into 11, adding 0s to the 11 for as long as you want to carry out the decimal:

$$
\begin{array}{r}
.647 \\
17\overline{)11.000} \\
10\,2 \\
\hline
80 \\
68 \\
\hline
120 \\
119 \\
\hline
1
\end{array}
$$

The final answer to three decimal places can then be expressed as 67.647. Using the same procedure, try converting an answer of $42\frac{16}{23}$ to a decimal. For most purposes, two digits past the decimal point is plenty.

$$
\begin{array}{r}
.69 \\
23\overline{)16.00} \\
13\,8 \\
\hline
2\,20 \\
2\,07 \\
\hline
13
\end{array}
$$

The final answer can then be expressed as 42.69.

As you might guess, I like to work some magic when I convert fractions to decimals. In the case of 1-digit fractions, the best way is to commit the following fractions—from halves through elevenths—to memory. This isn't as hard as it sounds. As you'll see below, most 1-digit fractions have special properties that make them hard to forget. Anytime you can reduce a fraction to one you already know, you'll speed up the process.

Chances are you already know the decimal equivalent of the following fractions:

$$\frac{1}{2} = .50 \qquad \frac{1}{3} = .\overline{333} \qquad \frac{2}{3} = .\overline{666}$$

(The bar signifies that the numbers will repeat indefinitely.)

Likewise:

$$\frac{1}{4} = .25 \qquad \frac{2}{4} = \frac{1}{2} = .50 \qquad \frac{3}{4} = .75$$

The 5ths are easy to remember:

$$\frac{1}{5} = .20 \qquad \frac{2}{5} = .40 \qquad \frac{3}{5} = .60 \qquad \frac{4}{5} = .80$$

The 6ths require memorizing only two new answers:

$$\frac{1}{6} = .1\overline{666} \qquad \frac{2}{6} = \frac{1}{3} = .\overline{333} \qquad \frac{3}{6} = \frac{1}{2} = .50$$

$$\frac{4}{6} = \frac{2}{3} = .\overline{666} \qquad \frac{5}{6} = .8\overline{333}$$

I'll return to the 7ths in a moment. The 8ths are a breeze:

$$\frac{1}{8} = .125 \qquad \frac{2}{8} = \frac{1}{4} = .25$$

$$\frac{3}{8} = .375 \left(3 \times \frac{1}{8} = 3 \times .125 = .375\right) \qquad \frac{4}{8} = \frac{1}{2} = .50$$

$$\frac{5}{8} = .625 \left(5 \times \frac{1}{8} = 5 \times .125 = .625\right) \qquad \frac{6}{8} = \frac{3}{4} = .75$$

$$\frac{7}{8} = .875 \left(7 \times \frac{1}{8} = 7 \times .125 = .875\right)$$

The 9ths have a magic all their own:

$$\frac{1}{9} = .\overline{111} \qquad \frac{2}{9} = .\overline{222} \qquad \frac{3}{9} = .\overline{333} \qquad \frac{4}{9} = .\overline{444}$$

$$\frac{5}{9} = .\overline{555} \qquad \frac{6}{9} = .\overline{666} \qquad \frac{7}{9} = .\overline{777} \qquad \frac{8}{9} = .\overline{888}$$

The 10ths you already know:

$$\frac{1}{10} = .10 \qquad \frac{2}{10} = .20 \qquad \frac{3}{10} = .30$$

$$\frac{4}{10} = .40 \qquad \frac{5}{10} = .50 \qquad \frac{6}{10} = .60$$

$$\frac{7}{10} = .70 \qquad \frac{8}{10} = .80 \qquad \frac{9}{10} = .90$$

For the 11ths, if you remember that $\frac{1}{11} = .\overline{0909}$, the rest is easy.

$$\frac{1}{11} = .\overline{0909} \qquad \frac{2}{11} = .\overline{1818} \, (2 \times .\overline{0909})$$

$$\frac{3}{11} = .\overline{2727} \, (3 \times .\overline{0909}) \qquad \frac{4}{11} = .\overline{3636} \qquad \frac{5}{11} = .\overline{4545}$$

$$\frac{6}{11} = .\overline{5454} \qquad \frac{7}{11} = .\overline{6363} \qquad \frac{8}{11} = .\overline{7272}$$

$$\frac{9}{11} = .\overline{8181} \qquad \frac{10}{11} = .\overline{9090}$$

The 7ths are truly remarkable. Once you memorize $\frac{1}{7} = .\overline{142857}$, you can get all the other 7ths without having to compute them:

$$\frac{1}{7} = .\overline{142857} \qquad \frac{2}{7} = .\overline{285714} \qquad \frac{3}{7} = .\overline{428571}$$

$$\frac{4}{7} = .\overline{571428} \qquad \frac{5}{7} = .\overline{714285} \qquad \frac{6}{7} = .\overline{857142}$$

Note that the same pattern of numbers repeats itself in each fraction. Only the starting point varies. You can figure out the starting point in a flash by roughly dividing the denominator into the numerator. In the case of $\frac{2}{7}$, 7 goes into 20 two times (with a remainder), so you know you should use the sequence that begins with 2, or $.\overline{285714}$. Similarly, when it comes to $\frac{3}{7}$, 7 goes into 30 four times (with a remainder), so you know you should use the sequence that begins with 4, or $.\overline{428571}$. Likewise with the rest.

You will have to calculate fractions higher than $\frac{10}{11}$ as you would any other division problem. However, keep your eyes peeled for ways of simplifying such problems. For example, you can simplify the fraction $\frac{18}{34}$ by dividing both numbers by 2, to reduce it to $\frac{9}{17}$, which is easier to compute.

If the denominator of the fraction is an even number, you can simplify the fraction by reducing it in half, even if the numerator is odd. For example:

$$\frac{9}{14} = \frac{4.5}{7}$$

Dividing the numerator and denominator in half reduces it to a 7ths fraction. Although the 7ths sequence previously shown doesn't provide the decimal for $\frac{4.5}{7}$, once you begin the calculation, the number you memorized will pop up:

$$\begin{array}{r} .\overline{6428571} \\ 7\overline{)4.5000} \\ \underline{4\,2} \\ 3 \end{array}$$

As you can see, you needn't work out the entire problem. Once you've reduced it to dividing 3 by 7, you can make a great impression on an audience by rattling off this long string of numbers almost instantly!

You can even put this trick to use in the middle of a problem. If your fraction is $\frac{3}{16}$, look what happens:

$$\begin{array}{r} .1 \\ 16\overline{)3.000} \\ \underline{1\,6} \\ 1\,4 \end{array}$$

Once the problem is reduced to $\frac{14}{16}$, you can further reduce it to $\frac{7}{8}$, which you know to be .875. Thus $\frac{3}{16}$ = .1875.

TESTING FOR DIVISIBILITY

In the last section, we saw how division problems could be simplified when both numbers are divisible by a common factor. We will end this chapter with a brief discussion on how to determine whether one number is a factor of another number. Being able to find the factors of a number helps us simplify division problems and, as we saw in Chapter 3, can speed up many multiplication problems. This will also be a very useful tool when we get to advanced multiplication in Chapter 8, as you will often be looking for ways to factor a 2-, 3-, or even a 5-digit number in the middle of a multiplication problem. Being able to factor these numbers quickly is very handy.

It's easy to test whether a number is divisible by 2. All you need to do is check if the last digit is even. If the last digit is 2, 4, 6, 8, or 0, the entire number is divisible by 2.

To test whether a number is divisible by 4, check the last two digits. The number 57,852 is a multiple of 4 because 52 = 13 × 4. The number 69,346 is not a multiple of 4 because to divide 4 into 46 will leave a remainder. The reason this works is because 4 divides evenly into 100 and thus into any multiple of 100. Thus, since 4 divides evenly into 57,800, and 4 divides into 52, 4 divides evenly into their sum of 57,852.

Likewise, since 8 divides into 1000, to test for divisibility by 8 check the last three digits of the number. For the number 14,918, divide 8 into 918. Since this leaves you with a remainder (918 ÷ 8 = $114\frac{6}{8}$), the number is not divisible by 8. You could also have observed this from the last two digits (18), which are not divisible by 4. Thus, 14,918 is not divisible by 4, so it can't be divisible by 8 either.

When it comes to divisibility by 3, here's a cool rule that's easy to remember: A number is divisible by 3 if and only if the sum of its digits are divisible by 3, no matter how many digits are in the number. To test whether 57,852 is divisible by 3, simply add 5 + 7 + 8 + 5 + 2 = 27. Since 27 is a multiple of 3, 57,852 is a multiple of 3. The same amazing rule holds true for divisibility by 9. A number is divisible by 9 if and only if its digits sum to a multiple of 9. Hence 57,852 is a multiple of 9, whereas 31,416 is not. The reason this works is based on the fact that the numbers 1, 10, 100, 1000, 10,000, and so on are all one greater than a multiple of 9.

To find out whether a number is divisible by 6, test whether the number is divisible by 2 and by 3. If it is, then the number must be divisible by 6.

Establishing whether a number is divisible by 5 is even easier. Any number, no matter how large, is a multiple of 5 if and only if it ends in either 5 or 0.

Establishing divisibility by 11 is almost as easy as determining divisibility by 3 or 9. A number is divisible by 11 if and only if you arrive at either 0 or a multiple of 11 when you alternately subtract and add the digits of the number. For instance, 73,958 is not divisible by 11, since $7 - 3 + 9 - 5 + 8 = 17$. However, the numbers 8492 and 73,194 are multiples of 11, since $8 - 4 + 9 - 2 = 11$ and $7 - 3 + 1 - 9 + 4 = 0$. The reason this works is based, like the rule for 3s and 9s, on the fact that the numbers 1, 100, 10,000, and 1,000,000 are one more than a multiple of 11, whereas the numbers 10, 1000, 100,000, and so on are one less than a multiple of 11.

Testing a number for divisibility by 7 is a bit trickier. If you add or subtract a number that is a multiple of 7 to the number you are testing and the resulting number is a multiple of 7, then the test is positive. I always choose to add or subtract a multiple of 7 so that the resulting sum or difference ends in 0. For example, to test the number 5292, I subtract 42 (a multiple of 7) to obtain 5250. Next, I get rid of the 0 at the end, leaving me with 525. Then I repeat the process by adding 35 (a multiple of 7) to 525, which gives me 560. When I delete the 0, I'm left with 56, which I know to be a multiple of 7. Therefore, 5292 is divisible by 7.

This method works not only for 7s, but for any odd number that doesn't end in 5. For example, to test whether 8792 is divisible by 13, subtract $4 \times 13 = 52$ from 8792 to arrive at 8740. Dropping the zero results in 874. Then add $2 \times 13 = 26$ to arrive at 900. Dropping the two zeros leaves you with 9, which clearly is not a multiple of 13. Therefore, 8792 is not a multiple of 13.

Exercise: 1-Digit and 2-Digit Division Exercises

Here you'll find a variety of 1-digit and 2-digit division problems that will test your mental prowess using the rule-of-thumb and simplification techniques explained earlier in this chapter. Check at the end of the book for answers and computations.

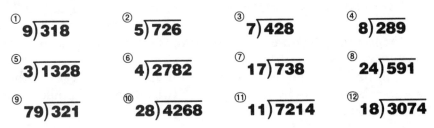

① $9\overline{)318}$ ② $5\overline{)726}$ ③ $7\overline{)428}$ ④ $8\overline{)289}$

⑤ $3\overline{)1328}$ ⑥ $4\overline{)2782}$ ⑦ $17\overline{)738}$ ⑧ $24\overline{)591}$

⑨ $79\overline{)321}$ ⑩ $28\overline{)4268}$ ⑪ $11\overline{)7214}$ ⑫ $18\overline{)3074}$

Decimalization Exercises

To solve the following problems, don't forget to employ the various 1-digit fractions you already know as decimals. Wherever appropriate, simplify the fraction before converting it to a decimal.

① $\dfrac{2}{5}$ ② $\dfrac{4}{7}$ ③ $\dfrac{3}{8}$ ④ $\dfrac{9}{12}$ ⑤ $\dfrac{5}{12}$ ⑥ $\dfrac{6}{11}$

⑦ $\dfrac{14}{24}$ ⑧ $\dfrac{13}{27}$ ⑨ $\dfrac{18}{48}$ ⑩ $\dfrac{10}{14}$ ⑪ $\dfrac{6}{32}$

Testing for Divisibility Exercises

In this final set of exercises, be especially careful testing for divisibility by 7. The rest should be easy for you.

DIVISIBILITY BY 2:

① 53428   ② 293   ③ 7241   ④ 9846

DIVISIBILITY BY 4:

⑤ 3932   ⑥ 67348   ⑦ 358   ⑧ 57929

DIVISIBILITY BY 8:

⑨ 59366   ⑩ 73488   ⑪ 248   ⑫ 6111

DIVISIBILITY BY 3:

⑬ 83671   ⑭ 94737   ⑮ 7359   ⑯ 3267486

DIVISIBILITY BY 6:

⑰ 5334   ⑱ 67386   ⑲ 248   ⑳ 5991

DIVISIBILITY BY 9:

㉑ 1234   ㉒ 8469   ㉓ 4425575   ㉔ 314159265

DIVISIBILITY BY 5:

㉕ 47830   ㉖ 43762   ㉗ 56785   ㉘ 37210

DIVISIBILITY BY 11:

^㉙53867   ^㉚4969   ^㉛3828   ^㉜941369

DIVISIBILITY BY 7:

^㉝5784   ^㉞7336   ^㉟875   ^㊱1183

DIVISIBILITY BY 17:

^㊲694   ^㊳629   ^㊴8273   ^㊵13855

The Art of "Guesstimation"

So far, you've been perfecting the mental techniques necessary to figure out the exact answers to math problems. Often, however, all you'll want is a ballpark estimate. Say you're getting quotes from different lenders on refinancing your home. All you really need at this information-gathering stage is a ballpark estimate of what your monthly payments will be. Or say you're settling a restaurant bill with a group of friends and you don't want to figure each person's bill to the penny. The guesstimation methods described in this chapter will make both these tasks—and many more just like them—much easier. Addition, subtraction, division, and multiplication all lend themselves to guesstimation. As usual, you'll do your computations from left to right.

Addition Guesstimation

Guesstimation is a good way to make your life easier when the numbers of a problem are too long to remember. The trick is to round the original numbers up or down:

$$
\begin{array}{r}
8{,}367 \\
+\ 5{,}819 \\
\hline
14{,}186
\end{array}
\approx
\begin{array}{r}
8{,}000 \\
6{,}000 \\
\hline
14{,}000
\end{array}
$$

($\approx$ means approximately)

Notice that we rounded the first number down to the nearest thousand and the second number up. Since the exact answer is 14,186, our relative error is only $\frac{186}{14{,}186}$, or 1.3%.

If you want to be more exact, instead of rounding off to the nearest thousand, round off to the nearest hundred:

$$
\begin{array}{r}
8{,}367 \\
+\ 5{,}819 \\
\hline
14{,}186
\end{array}
\quad\approx\quad
\begin{array}{r}
8{,}400 \\
5{,}800 \\
\hline
14{,}200
\end{array}
$$

The answer is only 14 off from the exact answer, an error of less than .1%. This is what I call a good guesstimation!

Try a 5-digit addition problem, rounding to the nearest hundred:

$$
\begin{array}{r}
46{,}187 \\
+\ 19{,}378 \\
\hline
65{,}565
\end{array}
\quad\approx\quad
\begin{array}{r}
46{,}200 \\
19{,}400 \\
\hline
65{,}600
\end{array}
$$

By rounding to the nearest hundred our answer will always be off by less than 100. If the answer is larger than 10,000, your guesstimate will be within 1%.

Now let's try something wild:

$$
\begin{array}{r}
23{,}859{,}379 \\
+\ 7{,}426{,}087 \\
\hline
31{,}285{,}466
\end{array}
\approx
\begin{array}{r}
24{,}000{,}000 \\
+\ 7{,}000{,}000 \\
\hline
31{,}000{,}000
\end{array}
\ \text{or}\
\begin{array}{r}
23.9 \text{ million} \\
+\ 7.4 \text{ million} \\
\hline
31.3 \text{ million}
\end{array}
$$

If you round to the nearest million, you get an answer of 31 million, off by roughly 285,000. Not bad, but you can do better by rounding to the nearest hundred thousand, as we've shown in the right-hand column. In this case you're off by only 15,000, which is awfully close when you're dealing with numbers in the millions. In general, if you round both numbers to the third digit of the larger number (here the hundred-thousand digit), your guesstimate will always be within 1% of the precise answer. If you can compute these smaller problems exactly, you can guesstimate the answer to any addition problem.

Guesstimating at the Supermarket

Let's try a real-world example. Have you ever gone to the store and wondered what the total is going to be before the cashier rings it up? One way to find out quickly, without a calculator or pencil and paper, is to total the prices mentally. My technique is to round the prices to the nearest 50¢. For example, while the cashier is adding the numbers shown below on the left, I mentally add the numbers shown on the right:

$ 1.39	$ 1.50
$ 0.87	$ 1.00
$ 2.46	$ 2.50
$ 0.61	$ 0.50
$ 3.29	$ 3.50
$ 2.99	$ 3.00
$ 0.20	$ 0.00
$ 1.17	$ 1.00
$ 0.65	$ 0.50
$ 2.93	$ 3.00
$ 3.19	$ 3.00
$19.75	$19.50

My final figure is usually within a dollar of the exact answer.

Subtraction Guesstimation

The way to guesstimate the answers to subtraction problems is the same—you round to the nearest thousand or hundreds digit, preferably the latter:

$$8{,}367 - 5{,}819 \approx 8{,}000 - 6{,}000 \quad \text{or} \quad 8{,}400 - 5{,}800$$
$$2{,}548 \qquad\qquad 2{,}000 \qquad\qquad 2{,}600$$

You can see that rounding to the nearest thousand leaves you with an answer quite a bit off the mark. By rounding to the *second* digit (hundreds, in the example), your answer will usually be within 3% of the exact answer. For this problem, your answer is off by only 52, a relative error of 2%. If you round to the third digit, the relative error will usually be below 1%. For instance:

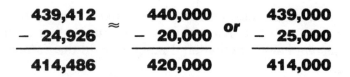

$$
\begin{array}{r}
439,412 \\
-\ 24,926 \\
\hline
414,486
\end{array}
\approx
\begin{array}{r}
440,000 \\
-\ 20,000 \\
\hline
420,000
\end{array}
\quad or \quad
\begin{array}{r}
439,000 \\
-\ 25,000 \\
\hline
414,000
\end{array}
$$

By rounding the numbers to the third digit rather than to the second digit, you improve the accuracy of the estimate by a significant amount. The first guesstimate is off by about 1.3%, whereas the second guesstimate is only off by about 0.16%.

Division Guesstimation

The first step in guesstimating the answer to a division problem is to determine the magnitude of the answer:

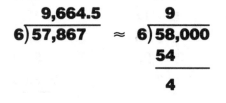

$$
\begin{array}{r}
9,664.5 \\
6\overline{)57,867}
\end{array}
\approx
\begin{array}{r}
9 \\
6\overline{)58,000} \\
54 \\
\hline
4
\end{array}
$$

Answer $= 9\frac{2}{3}$ **thousand** $= 9,667$

The next step is to round off the larger numbers to the nearest thousand and change the 57,867 to 58,000. Dividing 6 into 58 is simple. The answer is 9 with a remainder. But the most important component in this problem is where to place the 9.

For example, multiplying 6 × 90 yields 540, while multiplying 6 × 900 yields 5400, both of which are too small. But 6 × 9000 = 54,000, which is pretty close to the answer. This tells you the answer is 9 thousand and something. You can estimate just what that something is by first subtracting 58 − 54 = 4. At this point you could bring down the 0 and divide 6 into 40, and so forth. But if you're on your toes you'll realize that dividing 6 into 4 gives you $\frac{4}{6} = \frac{2}{3} = .667$. Since you know the answer is 9 thousand something, you're now in a position to guess 9667. In fact, the actual answer is 9644.5—darn close!

Division on this level is simple. But what about large division problems? Let's say we want to compute, just for fun, the amount of money a professional athlete earns a day if he makes $5,000,000 a year:

$$365 \text{ days}\overline{)\$5,000,000}$$

George Parker Bidder: The Calculating Engineer

The British have had their share of lightning calculators, and the mental performances of George Parker Bidder (1806–1878), born in Devonshire, were as impressive as any. Like most lightning calculators, Bidder began to try his hand (and mind) at mental arithmetic as a young lad. Learning to count, add, subtract, multiply, and divide by playing with marbles, Bidder went on tour with his father at age nine.

Almost no question was too difficult for him to handle. "If the moon is 123,256 miles from the earth and sound travels four miles a minute, how long would it take for sound to travel from the earth to the moon?" The young Bidder, his face wrinkled in thought for nearly a minute, replied "21 days, 9 hours, 34 minutes." (We know now that the distance is closer to 240,000 miles and sound cannot travel through the vacuum of space.) At age ten Bidder mentally computed the square root of 119,550,669,121 at 345,761 in a mere 30 seconds. In 1818 Bidder and the American lightning calculator Zerah Colburn (see page 36), were paired in a mental calculating duel in which Bidder, apparently, "outnumbered" Colburn.

Riding on his fame, George Bidder entered the University of Edinburgh and went on to become one of the more respected engineers in England. In parliamentary debates over railroad conflicts, Bidder was frequently called as a witness, which made the opposition shudder; as one said, "Nature had endowed him with particular qualities that did not place his opponents on a fair footing." Unlike Colburn, who retired as a lightning calculator at age 20, Bidder kept it up for his entire life. As late as 1878, in fact, just before his death, Bidder calculated the number of vibrations of light striking the eye in one second, based on the fact that there are 36,918 waves of red light per inch, and light travels at approximately 190,000 miles per second.

First you must determine the magnitude of the answer. Does this player earn thousands every day? Well, $365 \times 1000 = 365,000$, which is too low.

Does he earn tens of thousands every day? Well, $365 \times 10,000 = 3,650,000$, and that's more like it. To guesstimate your answer, divide the first two digits (or 36 into 50) and figure that's $1\frac{14}{36}$, or $1\frac{7}{18}$. Since 18 goes into 70 about 4 times, your guess is that the athlete earns about $14,000 per day. The exact answer is $13,698.63 per day. Not a bad estimate (and not a bad salary!).

Here's an *astronomical* calculation for you. How many seconds does it take light to get from the Sun to the Earth? Well, light travels at 186,282 miles per second, and the sun is (on average) 92,960,130 miles away:

$$186,282\overline{)92,960,130}$$

I doubt you're particularly eager to attempt this problem by hand. Fortunately, it's relatively simple to guesstimate an answer. First, simplify the problem:

$$\approx 186,000\overline{)93,000,000} = 186\overline{)93,000}$$

Now divide 186 into 930, which yields 5 with no remainder. Then tack back on the two zeros you removed from 93,000 and voilà—your answer is 500 seconds. The exact answer is 499.02 seconds, so this is a pretty respectable guesstimate.

Multiplication Guesstimation

Remember 2-digit multiplication from Chapter 3? You can use much the same techniques to guesstimate your answers. For example:

$$
\begin{array}{r} 88 \\ \times\ 54 \\ \hline 4752 \end{array}
\quad\approx\quad
\begin{array}{r} 90 \\ \times\ 50 \\ \hline 4500 \end{array}
$$

Rounding up to the nearest multiple of 10 simplifies the problem considerably, but you're still off by 252, or about 5%. You can do better if you round both numbers by the same amount. That is, if you round 88 by increasing 2, you should also decrease 54 by 2:

$$
\begin{array}{r} 88 \\ \times\ 54 \\ \hline 4752 \end{array}
\quad\approx\quad
\begin{array}{r} 90 \\ \times\ 52 \\ \hline 4680 \end{array}
$$

Instead of a 1-by-1 multiplication problem you now have a 2-by-1 problem, which should be easy enough for you to do. Your guesstimate is off by only 1.5%.

You'll notice when you guesstimate the answer to multiplication problems that if you round the top number up and the bottom number down so that the numbers are farther apart, your guesstimate will be a little low. If you round the top number down and the bottom number up so that the numbers are closer together, your guesstimate will be a little high. The larger the amount by which you round up or down, the greater your guesstimate will be off from the exact answer. For example:

$$
\begin{array}{cc}
73 & 70 \\
\times\,65 & \times\,68 \\
\hline
4745 & 4760
\end{array}
\qquad \approx
$$

Since the numbers are closer together after you round them off, your guesstimate is a little high. The opposite is true when you round off the numbers so that they're farther apart:

$$
\begin{array}{cc}
67 & 70 \\
\times\,67 & \times\,64 \\
\hline
4489 & 4480
\end{array}
\qquad \approx
$$

Since the numbers are farther apart, the estimated answer is too low, though again, not by much. You can see that this multiplication guesstimation method works quite well. Also notice that this problem is just 67^2 and that our approximation is just the first step of the squaring techniques of Chapter 2. Let's look at one more example:

$$
\begin{array}{cc}
83 & 85 \\
\times\,52 & \times\,50 \\
\hline
4316 & 4250
\end{array}
\qquad \approx
$$

We observe that the approximation is most accurate when the original numbers are close together. In Chapter 8, where we take up advanced multiplication, we will go into the theory behind this method and explain how to go from the approximation to an exact answer.

Try estimating a 3 × 2 multiplication problem:

$$
\begin{array}{cc}
728 & 731 \\
\times63 & \times60 \\
\hline
45{,}864 & 43{,}860
\end{array}
\qquad \approx
$$

By rounding 63 down to 60 and 728 up to 731, you create a 3-by-1 multiplication problem, which puts your guesstimate within 2004 of the exact answer, an error of 4.3%. To be more accurate you need to be proficient at doing 2-by-2 problems. For instance, you could round 728 to 730:

$$
\begin{array}{r}
728 \\
\times\ \ \ 63 \\
\hline
45{,}864
\end{array}
\quad\approx\quad
\begin{array}{r}
730 \\
\times\ \ \ 63 \\
\hline
45{,}990
\end{array}
$$

Since $73 \times 63 = 73 \times 9 \times 7 = 657 \times 7 = 4599$, $730 \times 63 = 45{,}990$, a much closer estimate. In this case your guesstimate is too high by $2 \times 63 = 126$, or 0.3%.

Now try guesstimating the following 3-by-3 problem:

$$
\begin{array}{r}
367 \\
\times\ \ \ 492 \\
\hline
180{,}564
\end{array}
\quad\approx\quad
\begin{array}{r}
359 \\
\times\ \ \ 500 \\
\hline
179{,}500
\end{array}
$$

You will notice that although you rounded both numbers up and down by 8, your guesstimate is off by over 1000. That's because the multiplication problem is larger and the size of the rounding number is larger, so the resulting estimate will be off by a greater amount. But the relative error is still under 1%.

How high can you go with this system of guesstimating multiplication problems? As high as you want. You just need to know the names of large numbers. A *thousand thousand* is a *million*, and a *thousand million* is a *billion*. Knowing these names and numbers, try this one on for size:

$$
\begin{array}{r}
28{,}657{,}493 \\
\times\ \ \ \ \ \ \ \ 13{,}864 \\
\hline
\end{array}
\qquad
\begin{array}{r}
29\ \text{million} \\
\times\ 14\ \text{thousand} \\
\hline
\end{array}
$$

As before, the objective is to round the numbers to simpler numbers such as 29,000,000 and 14,000. Dropping the 0s for now, this is just a 2-by-2 multiplication problem: $29 \times 14 = 406$. Hence the answer is roughly 406 billion, since a thousand million is a billion.

SQUARE ROOT ESTIMATION: DIVIDE AND AVERAGE

The square root of a number, x, is the number which, when multiplied by itself, will give you x. For example, the square root of 9 is 3 because 3 × 3 = 9. The square root is used in many science and engineering problems and is almost always solved with a calculator. The following method provides an accurate estimate of the answer.

In square root estimation your goal is to come up with a number that when multiplied by itself approximates the original number. Since the square root of most numbers is not a whole number, your estimate is likely to contain a fraction or decimal point.

Let's start by guesstimating the square root of 19. Your first step is to think of the number that when multiplied by itself comes closest to 19. Well, 4 × 4 = 16 and 5 × 5 = 25. Since 25 is too high, the answer must be 4 point something. Your next step is to divide 4 into 19, giving you 4.75. Now, since 4 × 4 is less than 4 × 4.75 = 19, which in turn is less than 4.75 × 4.75, we know that 19 (or 4 × 4.75) lies between 4² and (4.75)². Hence, the square root of 19 lies between 4 and 4.75.

I'd guess the square root of 19 to be about halfway between, at 4.375. In fact, the square root of 19 (rounded to 3 decimal places) is 4.359, so our guesstimate is pretty close. We illustrate this procedure as follows:

Divide:

$$
\begin{array}{r}
4.75 \\
4\overline{)19.0} \\
16 \\
\hline
3\ 0 \\
2\ 8 \\
\hline
20 \\
20 \\
\hline
0
\end{array}
$$

Average:

$$\frac{4 + 4.75}{2} = 4.375$$

Now you try a slightly harder one. What's the square root of 87?

Divide:	**Average:**
9.66	
$9\overline{)87.0}$	$\dfrac{9 + 9.66}{2} = 9.33$

First come up with your ballpark figure, which you can get fairly quickly by noting that $9 \times 9 = 81$ and $10 \times 10 = 100$, which means the answer is 9 point something. Carrying out the division of 9 into 87 to two decimal places, you get 9.66. To improve your guesstimate, take the average of 9 and 9.66, which is 9.33—exactly the square root of 87 rounded to the second decimal place!

Let's face it, it's pretty easy to guesstimate the square root of 2-digit numbers. But what about 3-digit numbers? Actually, they're not much harder. I can tell you right off the bat that all 3-digit and 4-digit numbers have 2-digit square roots before the decimal point. And the procedure for computing square roots is the same, no matter how large the number. For instance, to compute the square root of 679, first find your ballpark figure. Because 20 squared is 400 and 30 squared is 900, the square root of 679 must lie between 20 and 30.

When you divide 20 into 679, stop at your first two digits, which are 33. Averaging 20 and 33 gives you your final guesstimate of 26.5. In fact, the exact square root is 26.05.

Divide:	**Average:**
33	
$20\overline{)679}$	$\dfrac{20 + 33}{2} = 26.5$

There are two reasons why I average 20 with 33 instead of with 33.95, the exact quotient. First of all, it takes less time to arrive at 33. Second, if you take your average with the exact quotient, you'll always overestimate the square root. By stopping at 33 you bring your average down a little.

To guesstimate the square root of 4-digit numbers, look at the first two digits of the number to determine the first digit of the square root. For example, to find the square root of 7369 consider the square root of 73. Since $8 \times 8 = 64$ and $9 \times 9 = 81$, 8 must be the first digit of the square root. So the answer is 80 something. Now proceed the usual way:

Divide: **Average:**

$$80\overline{)7369} \quad \frac{92}{}$$

$$\frac{80 + 92}{2} = 86$$

The exact answer (to 3 places) is 85.8!

To guesstimate the square root of a 6-digit number like 593,472 would seem like an impossible task for the uninitiated, but for you it's no sweat. Since $700^2 = 490,000$, and $800^2 = 640,000$, the square of 593,472 must lie between 700 and 800. In fact, all 5-digit and 6-digit numbers have 3-digit square roots. In practice, you only need to look at the square root of the first two digits of 6-digit numbers (or the first digit of 5-digit numbers). Once you figure out that the square root of 59 lies between 7 and 8, you know your answer is in the 700s.

Now proceed in the usual manner:

Divide: **Average:**

$$700\overline{)593472}$$

$$847$$
$$\approx 7\overline{)5934}$$

$$\frac{700 + 847}{2} = 773.5$$

The exact square root of 593,472 is 770.47 (to five places) so you're pretty close. But you could have been closer, as the following trick demonstrates. Note that the first two digits, 59, are closer to 64 (8 × 8) than they are to 49 (7 × 7). Because of this you can start your guesstimation with the number 8 and proceed from there:

Divide: **Average:**

$$800\overline{)593472}$$

$$741$$
$$\approx 8\overline{)5934}$$

$$\frac{800 + 741}{2} = 770.5$$

Just for fun, let's do a real whopper—the square root of 28,674,529. This isn't as hard as it might seem. Your first step is to round to the nearest large number—in this case, just find the square root of 29.

The Mathematical Duel of Évariste Galois

The tragic story of the French mathematician Évariste Galois (1812–1832), killed at the age of 20 in a duel over "an infamous coquette," is legendary in the annals of the history of mathematics. A precociously brilliant student, Galois lay the foundation for a branch of mathematics known as group theory. Legend has it that he penned his theory the night before the duel, anticipating his demise and wanting to leave his legacy to the mathematics community. Hours before his death, on May 30, 1832, Galois wrote to Auguste Chevalier: "I have made some new discoveries in analysis. The first concerns the theory of equations, the others integral functions." After describing these he asked his friend: "Make a public request of Jacobi or Gauss to give their opinions not as to the truth but as to the importance of these theorems. After that, I hope some men will find it profitable to sort out this mess."

Romantic legend and historical truth, however, do not always match. What Galois penned the night before his death were corrections and editorial changes to papers that had been accepted by the Academy of Sciences long before. Further, Galois's initial papers had been submitted three years prior to the duel, when he was all of 17! It was after this that Galois became embroiled in political controversy, was arrested, spent time in a prison dungeon and, ultimately, got himself mixed up in a dispute over a woman and killed.

Aware of his own precocity Galois noted: "I have carried out researches which will halt many savants in theirs." For over a century that proved to be the case.

Divide:

$$5) \overline{29.0} \quad \begin{array}{r} 5.8 \\ \end{array}$$

$$\begin{array}{r} 5.8 \\ 5)\overline{29.0} \\ 25 \\ \hline 4\ 0 \end{array}$$

Average:

$$\frac{5 + 5.8}{2} = 5.4$$

All 7-digit and 8-digit numbers have 4-digit square roots, so 5.4 becomes 5400, your final estimate. The exact answer is 5354.8. Not bad!

This wraps up the chapter on guesstimation math. After doing the exercises below, turn to the next chapter on pencil-and-paper math where you will learn to generate exact answers for many of these guesstimation problems, but in a much quicker way than you've done on paper before.

GUESSTIMATION EXERCISES

Go through the following exercises for guesstimation math; then check your answers and computations with ours at the back of the book.

Addition Guesstimation Exercises

Round these numbers up or down and see how close you can come to the exact answer:

①	②	③	④
1,479	57,293	312,025	8,971,011
+ 1,105	+ 37,421	+ 79,419	+ 4,016,367

Mentally estimate the total for the following column of price numbers by rounding to the nearest 50¢:

$ 2.67
$ 1.95
$ 7.35
$ 9.21
$ 0.49
$11.21
$ 0.12
$ 6.14
$ 8.31

Subtraction Guesstimation Exercises

Estimate the following subtraction problems by rounding to the second or third digit:

①	②	③	④
4926	67,221	526,978	8,349,241
− 1659	− 9,874	− 42,009	− 6,103,839

Division Guesstimation Exercises

Adjust the numbers in a way that allows you to guesstimate the following division problems:

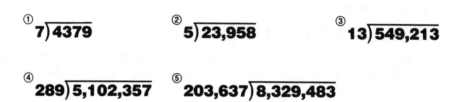

① 7)4379   ② 5)23,958   ③ 13)549,213

④ 289)5,102,357   ⑤ 203,637)8,329,483

Multiplication Guesstimation Exercises

Adjust the numbers in a way that allows you to guesstimate the following multiplication problems:

① 98
× 27

② 76
× 42

③ 88
× 88

④ 539
× 17

⑤ 312
× 98

⑥ 639
× 107

⑦ 428
× 313

⑧ 51,276
× 489

⑨ 104,972
× 11,201

⑩ 5,462,741
× 203,413

Square Root Guesstimation Exercises

Estimate the square roots of the following numbers using the *divide and average* method:

① $\sqrt{17}$ ② $\sqrt{35}$ ③ $\sqrt{163}$ ④ $\sqrt{4279}$ ⑤ $\sqrt{8039}$

6

Math for the Board: Pencil-and-Paper Mathematics

In the preface to this book I discussed the many benefits you will get from being able to do mental calculations. In this chapter I present some methods for speeding up pencil-and-paper calculations as well. Since calculators have replaced much of the need for pencil-and-paper arithmetic in most practical situations, I've chosen to concentrate on the lost art of calculating square roots and the flashy criss-cross method for multiplying large numbers. Since these are, admittedly, mostly for mental gymnastics and not for some practical application, I will first touch on addition and subtraction and show you just a couple of little tricks for speeding up the process and for checking your answers. These techniques *can* be used in daily life, as you'll see.

If you are anxious to get to the more challenging multiplication problems, you can skip this chapter and go directly to Chapter 7, which *is* critical for mastering the big problems in Chapter 8. If you need a break and just want to have some fun, then I recommend going through this chapter—you'll enjoy playing with pencil and paper once again.

COLUMNS OF NUMBERS

Adding long columns of numbers is just the sort of problem you might run into in business or while figuring out your personal finances. Add the following column of numbers as you normally would and then check to see how I do it.

$$
\begin{array}{r}
4328 \\
884 \\
620 \\
1477 \\
617 \\
+\ \ 725 \\
\hline
8651
\end{array}
$$

When I have pencil and paper at my disposal, I add the numbers from top to bottom and from right to left, just as I learned to do in school. As I sum the digits, the only numbers that I "hear" are the partial sums. That is, when I sum the first (rightmost) column 8 + 4 + 0 + 7 + 7 + 5, I hear "8 . . . 12 . . . 19 . . . 26 . . . 31." Then I put down the 1, carry the 3, and proceed as usual. Once I have my final answer, I check my computation by adding the numbers from bottom to top and, I hope, arrive at the same answer.

For instance, the first column would be summed as 5 + 7 + 7 + 0 + 4 + 8 (which, in my mind, sounds like "5 . . . 12 . . . 19 . . . 23 . . . 31"). Then I put down the 1 and add 3 + 2 + 1 + 7 + 2 + 8 + 2, and so on. By adding the numbers in a different order, you are less likely to make the same mistake twice. Of course, if the answers differ, then at least one of the calculations must be wrong.

MOD SUMS

If I'm feeling frisky, I check my answer by a method I call "mod sums" (because it is based on the elegant mathematics of modular arithmetic). I admit this method is not as practical, but it's easy to use and is the basis for some amazing mathemagical tricks that we will do in Chapter 9.

With the mod sums method you sum the digits of each number until you are left with a single digit. For example, to compute the mod sum of 4328, add 4 + 3 + 2 + 8 = 17. Then add the digits of 17 to get 1 + 7 = 8. Hence the mod sum of 4328 is 8. For the previous problem the mod sums of each number are computed as follows:

```
4328  →  17  →                    8
 884  →  20  →                    2
 620  →   8  →                    8
1477  →  19  →  10  →             1
 617  →  14  →                    5
+ 725  →  14  →                 +  5
─────                           ────
8651                             29
  ⊥                               ⊥
 20                              11
  ⊥                               ⊥
  ②                               ②
```

As illustrated above, the next step is to add all the mod sums together (8 + 2 + 8 + 1 + 5 + 5). This yields 29, which sums to 11, which in turn sums to 2. Note that the mod sum of 8651, your original total of the original digits, is also 2. This is not a coincidence! If you've computed the answer and the mod sums correctly, your final mod sums must be the same. If they are different, you have definitely made a mistake somewhere: There is a 1 in 9 chance that the mod sums will match accidentally. If there is a mistake, then this method will detect it 8 times out of 9.

The mod sum method is more commonly known to mathematicians and accountants as "casting out 9s" because the mod sum of a number happens to be equal to the remainder obtained when the number is divided by 9. In the case of the answer above—8651—the mod sum was 2. If you divide 8651 by 9 the answer is 961 with a remainder of 2. In other words, if you cast out 9 from 8651 a total of 961 times, you'll have a remainder of 2. There's one small exception to this. Recall from Chapter 4 that the sum of the digits of any multiple of 9 is also a multiple of 9. Thus, if a number is a multiple of 9 it will have a mod sum of 9, even though it has a remainder of 0.

SUBTRACTING ON PAPER

You can't, of course, subtract columns of numbers the same way you add them. Rather, you subtract them number by number, which means that all subtraction problems involve just two numbers. Once again, with pencil and paper at our disposal it's easiest to subtract from right to left. To check your answer, just add the answer to the second number. If you are correct you should get the top number.

If you want you can also use mod sums to check your answer. The key is to *subtract* the mod sums you arrive at and then compare that number to the mod sum of your answer:

$$
\begin{array}{r}
65{,}717 \rightarrow \quad 8 \\
- \ 38{,}491 \rightarrow \ -7 \\
\hline
27{,}226 \rightarrow \quad ① \\
\end{array}
$$

There's one extra twist. If the difference in the mod sums is a negative number or 0, add 9 to it. For instance:

$$
\begin{array}{r}
42{,}689 \rightarrow \quad 2 \\
- \ 18{,}764 \rightarrow \ -8 \\
\hline
23{,}925 \qquad -6 + 9 = ③ \\
\end{array}
$$

$$\downarrow$$

$$21$$

$$③$$

PENCIL-AND-PAPER SQUARE ROOTS

With the advent of pocket calculators, the pencil-and-paper method of calculating square roots has practically become a lost art. You've already seen how to estimate square roots mentally. Now I'll show you how to do it exactly, using pencil and paper.

Remember in guesstimating square roots you calculated the square root of 19? Let's look at that problem again, this time using a method that will give you the exact square root.

$$
\begin{array}{r}
\textbf{4. 3 5 8} \\
\sqrt{\textbf{19.000000}} \\
\textbf{4}^2 = \textbf{16}
\end{array}
$$

$$
\begin{array}{r}
\textbf{8 } \times \underline{\ } \ \leq \ \textbf{3 00} \\
\textbf{83} \times \textbf{3} = \ \textbf{2 49}
\end{array}
$$

$$
\begin{array}{r}
\textbf{86 } \times \underline{\ } \ \leq \ \textbf{5100} \\
\textbf{865} \times \textbf{5} = \ \textbf{4325}
\end{array}
$$

$$
\begin{array}{r}
\textbf{870 } \times \underline{\ } \ \leq \ \textbf{77500} \\
\textbf{8708} \times \textbf{8} = \ \textbf{69664}
\end{array}
$$

I will describe the general method that fits all cases, and illustrate it with the above example.

STEP 1. If the number of digits to the left of the decimal point is one, three, five, seven, or any odd number of digits, the first digit of the answer (or quotient) will be the largest number whose square is less than the *first* digit of the original number. If the number of digits to the left of the decimal point is two, four, six, or any even number of digits, the first digit of the quotient will be the largest number whose square is less than the first *two* digits of the dividend. In this case, 19 is a 2-digit number, so the first digit of the quotient is the largest number whose square is less than 19. That number is 4. Write the answer above either the first digit of the dividend (if odd) or the second digit of the dividend (if even).

STEP 2. Subtract the square of the number in Step 1, then bring down two more digits. The number 4^2 is 16. Subtract $19 - 16 = 3$. Bring down two 0s, leaving 300 as the current remainder.

STEP 3. Double the current quotient (ignoring any decimal point), and put a blank space in front of it. Here $4 \times 2 = 8$. Put down 8_ × _ to the left of the current remainder, in this case 300.

STEP 4. The next digit of the quotient will be the largest number that can be put in both blanks so that the resulting multiplication problem is less than or equal to the current remainder. In this case the number is 3, because $83 \times 3 = 249$, whereas $84 \times 4 = 336$ is too high. Write this number above the second digit of the next two numbers; in this case the 3 would go above the second 0. We now have a quotient of 4.3.

STEP 5. If you want more digits, subtract the product from the remainder (i.e., $300 - 249 = 51$), and bring down the next two digits, in this case 51 turns into 5100, which becomes the current remainder. Now repeat Steps 3 and 4.

To get the third digit of the square root, double the quotient again ignoring the decimal point (i.e., $43 \times 2 = 86$). Place 86_ × _ to the left of 5100. The number 5 gives us $865 \times 5 = 4325$, the largest product below 5100. The 5 goes above the second digits of the next two numbers, in this case two more 0s. We now have a quotient of 4.35. For even more digits, repeat the process as we've done in the example.

Here's an example of an odd number of digits before the decimal point:

$$
\begin{array}{r}
\mathbf{2\ 8.\ 9\ 7} \\
\sqrt{\mathbf{839.4000}}
\end{array}
$$

$$2^2 = 4$$

$$
\begin{array}{rcr}
\mathbf{4}\_ \times \_ & \leq & \mathbf{439} \\
\mathbf{48} \times \mathbf{8} & = & \mathbf{384} \\
\hline
\mathbf{56}\_ \times \_ & \leq & \mathbf{55\ 40} \\
\mathbf{569} \times \mathbf{9} & = & \mathbf{51\ 21} \\
\hline
\mathbf{578}\_ \times \_ & \leq & \mathbf{4\ 1900} \\
\mathbf{5787} \times \mathbf{7} & = & \mathbf{4\ 0509} \\
\hline
\end{array}
$$

Next, we'll calculate the square root of a 4-digit number. In this case—as with 2-digit numbers—we consider the first two digits of the problem to determine the first digit of the square root:

$$
\begin{array}{r}
\mathbf{8\ 2.\ 0\ 6} \\
\sqrt{\mathbf{6735.0000}}
\end{array}
$$

$$8^2 = 64$$

$$
\begin{array}{rcr}
\mathbf{16}\_ \times \_ & \leq & \mathbf{335} \\
\mathbf{162} \times \mathbf{2} & = & \mathbf{324} \\
\hline
\mathbf{164}\_ \times \_ & \leq & \mathbf{11\ 00} \\
\mathbf{1640} \times \mathbf{0} & = & \mathbf{0} \\
\hline
\mathbf{1640}\_ \times \_ & \leq & \mathbf{11\ 0000} \\
\mathbf{16406} \times \mathbf{6} & = & \mathbf{9\ 8436} \\
\hline
\end{array}
$$

Finally, if the number you are calculating the square root for is a perfect square, you will know it as soon as you end up with a remainder of 0 and nothing left to bring down. For example:

$$
\begin{array}{r}
\textbf{3.3} \\
\sqrt{\textbf{10.89}} \\
3^2 = \quad 9 \\
\hline
6\_ \times \_ \leq \ \textbf{1 89} \\
6\underline{3} \times \underline{3} = \ \textbf{1 89} \\
\hline
\textbf{0}
\end{array}
$$

PENCIL-AND-PAPER MULTIPLICATION

For pencil-and-paper multiplication I use the "criss-cross method," which enables me to write down the entire answer on one line without ever writing any partial results! This is one of the most impressive displays of mathemagics when you have pencil and paper (or blackboard and chalk) at your disposal. Many lightning calculators in the past earned their reputations with this method. They would be given two large numbers and write down the answer almost instantaneously. Here's how it works:

$$
\begin{array}{r}
\textbf{47} \\
\times \quad \textbf{34} \\
\hline
\textbf{1598}
\end{array}
$$

The criss-cross method is best learned by example:

STEP 1. First, multiply 4 × 7 to yield 2<u>8</u>, put down the <u>8</u>, and carry the 2 to the next computation, below.

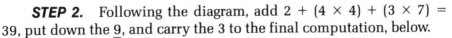

STEP 2. Following the diagram, add 2 + (4 × 4) + (3 × 7) = 3<u>9</u>, put down the <u>9</u>, and carry the 3 to the final computation, below.

STEP 3. End by adding 3 + (3 × 4) = <u>15</u> and put down <u>15</u> to arrive at your final answer.

$$4 \qquad 7$$
$$| \qquad$$
$$3 \qquad 4$$

Let's solve another 2-by-2 multiplication problem using the criss-cross method:

$$\begin{array}{r} 83 \\ \times \quad 65 \\ \hline 5395 \end{array}$$

The steps and diagrams appear as follows:

STEP 1. 5 × 3 = 1<u>5</u>.   8   3
                                                 |
                                          6   5

STEP 2. 1 + (5 × 8) + (6 × 3) = 5<u>9</u>.   8＼ 3
                                                                  ✕
                                                             6／ 5

STEP 3. 5 + (6 × 8) = <u>53</u>.   8   3
                                                     |
                                              6   5

The criss-cross method gets slightly more complicated with 3-by-3 problems:

$$\begin{array}{r} 853 \\ \times \quad 762 \\ \hline 649{,}986 \end{array}$$

We proceed as suggested by our pattern below:

STEP 1. 2 × 3 = <u>6</u>.   8   5   3
                                                       |
                                       7   6   2

STEP 2. (2 × 5) + (6 × 3) = 2<u>8</u>.   8   5＼ 3
                                                                ✕
                                                     7   6／ 2

STEP 3. $2 + (2 \times 8) + (7 \times 3) + (6 \times 5) = 6\underline{9}$.

STEP 4. $6 + (6 \times 8) + (5 \times 7) = 8\underline{9}$.

8   5   3

7   6   2

STEP 5. $8 + (8 \times 7) = \underline{64}$.

8   5   3

7   6   2

Notice that the number of multiplications at each step is 1, 2, 3, 2, and 1 respectively.

The mathematics underlying the criss-cross method is nothing more than the distributive law. For instance, $853 \times 762 = (800 + 50 + 3) \times (700 + 60 + 2)$, $= (3 \times 2) + [(5 \times 2) + (3 \times 6)] \times 10 + [(8 \times 2) + (5 \times 6) + (3 \times 7)] \times 100 + [(8 \times 6) + (5 \times 7)] \times 1000 + [8 \times 7] \times 10,000$, which are precisely the calculations of the criss-cross method.

You can check your answer with the mod sum method by multiplying the mod sums of the two numbers and computing the resulting number's mod sum. Compare this number to the mod sum of the answer. They should match. For example:

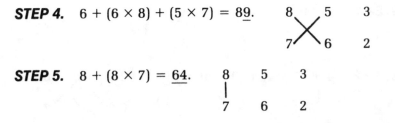

If the mod sums don't match, you made a mistake. This method will detect mistakes, on average, 8 times out of 9.

In the case of 3-by-2 multiplication problems, the procedure is the same except you treat the hundreds digit of the second number as a 0:

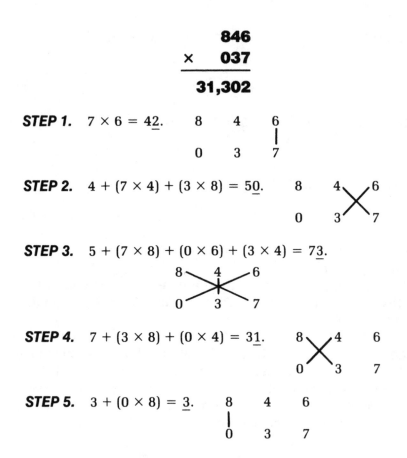

$$846$$
$$\times \quad 037$$
$$\overline{\quad 31,302 \quad}$$

STEP 1. $7 \times 6 = 4\underline{2}$. 8   4   6
                                    │
                                 0   3   7

STEP 2. $4 + (7 \times 4) + (3 \times 8) = 5\underline{0}$.   8   4   6
                                                        0   3   7

STEP 3. $5 + (7 \times 8) + (0 \times 6) + (3 \times 4) = 7\underline{3}$.
                                          8   4   6
                                        0   3   7

STEP 4. $7 + (3 \times 8) + (0 \times 4) = 3\underline{1}$.   8   4   6
                                                     0   3   7

STEP 5. $3 + (0 \times 8) = \underline{3}$.   8   4   6
                                                0   3   7

You can use the criss-cross to do any size multiplication problem. To answer the 5-by-5 problem below will require nine steps. The number of multiplications in each step is 1, 2, 3, 4, 5, 4, 3, 2, 1.

$$42,867$$
$$\times \quad 52,049$$
$$\overline{\quad 2,231,184,483 \quad}$$

STEP 1. $9 \times 7 = 6\underline{3}$.   4   2   8   6   7
                                          │
                                       5   2   0   4   9

STEP 2. $6 + (9 \times 6) + (4 \times 7) = 8\underline{8}$.
                                  4   2   8   6   7
                                5   2   0   4   9

STEP 3. $8 + (9 \times 8) + (0 \times 7) + (4 \times 6) = 10\underline{4}$.

$$\begin{array}{ccccc} 4 & 2 & 8 & 6 & 7 \\ 5 & 2 & 0 & 4 & 9 \end{array}$$

STEP 4. $10 + (9 \times 2) + (2 \times 7) + (4 \times 8) + (0 \times 6) = 7\underline{4}$.

$$\begin{array}{ccccc} 4 & 2 & 8 & 6 & 7 \\ 5 & 2 & 0 & 4 & 9 \end{array}$$

STEP 5. $7 + (9 \times 4) + (5 \times 7) + (4 \times 2) + (2 \times 6) + (0 \times 8) = 9\underline{8}$.

$$\begin{array}{ccccc} 4 & 2 & 8 & 6 & 7 \\ 5 & 2 & 0 & 4 & 9 \end{array}$$

STEP 6. $9 + (4 \times 4) + (5 \times 6) + (0 \times 2) + (2 \times 8) = 7\underline{1}$.

$$\begin{array}{ccccc} 4 & 2 & 8 & 6 & 7 \\ 5 & 2 & 0 & 4 & 9 \end{array}$$

STEP 7. $7 + (0 \times 4) + (5 \times 8) + (2 \times 2) = 5\underline{1}$.

$$\begin{array}{ccccc} 4 & 2 & 8 & 6 & 7 \\ 5 & 2 & 0 & 4 & 9 \end{array}$$

STEP 8. $5 + (2 \times 4) + (5 \times 2) = 2\underline{3}$.

$$\begin{array}{ccccc} 4 & 2 & 8 & 6 & 7 \\ 5 & 2 & 0 & 4 & 9 \end{array}$$

STEP 9. $(5 \times 4) + 2 = \underline{\underline{22}}$.

$$\begin{array}{ccccc} 4 & 2 & 8 & 6 & 7 \\ 5 & 2 & 0 & 4 & 9 \end{array}$$

You can check your answer by using the mod sum method:

$$\begin{array}{rcl} 42{,}867 & \rightarrow & 9 \\ \times \quad 52{,}049 & \rightarrow & \times\ 1 \\ \hline 2{,}231{,}184{,}483 & & 9 \\ \downarrow & & \downarrow \\ 36 & & ⑨ \\ \downarrow & & \\ ⑨ & & \end{array}$$

Shakuntala Devi: That's Incalculable!

In 1976 *The New York Times* reported that an Indian woman named Shakuntala Devi (b. 1940–   ) added 25,842 + 111,201,721 + 370,247,830 + 55,511,315, and then multiplied that sum by 9,878, for a correct total of 5,559,369,456,432, all in less than 20 seconds. Hard to believe, though this uneducated daughter of impoverished parents has made a name for herself in the United States and Europe as a lightning calculator.

Unfortunately, most of Devi's truly amazing feats not done by obvious "tricks of the trade" are poorly documented. Her greatest-claimed accomplishment—the timed multiplication of two 13-digit numbers on paper—still stands in the *Guinness Book of World Records* as an example of a "Human Computer." The time of the calculation, however, is questionable at best. Devi, a master of the criss-cross method (discussed in this chapter), reportedly multiplied 7,686,369,774,870 × 2,465,099,745,779, numbers randomly generated at the Computer Department of Imperial College, London, on June 18, 1980. Her correct answer of 18,947,668,177,995,426,773,730 was allegedly generated in an incredible 28 seconds. *Guinness* offers this disclaimer: "Some eminent mathematical writers have questioned the conditions under which this was apparently achieved and predict that it would be impossible for her to replicate such a feat under highly rigorous surveillance." The basic time needed to calculate 169 multiplication problems and 167 addition problems, for a total of 336 operations, would each have to be done in under a tenth of a second, with no mistakes, taking the time to write down all 26 digits of the answer. Reaction time alone makes this record truly in the category of "that's incalculable!"

Despite this, Devi has proven her abilities doing rapid calculation by many of the techniques in this book and has, in fact, written her own book on the subject.

CASTING OUT ELEVENS

To double-check your answer another way, you can use the method known as "casting out elevens." It's similar to casting out nines, except you reduce the numbers by alternately subtracting and adding the digits from right to left, ignoring any decimal point. If the result is negative, add eleven to it. (It may be tempting to do the addition and subtraction from left to right, as you do with mod sums, but in this case you must do it right to left or it will not always work.)

For example:

$$234.87 \dashrightarrow 7 - 8 + 4 - 3 + 2 = \quad 2 \dashrightarrow 2$$
$$+ \; 58.61 \dashrightarrow 1 - 6 + 8 - 5 \quad\quad = -2 \dashrightarrow 9$$
$$293.48 \dashrightarrow 8 - 4 + 3 - 9 + 2 = \textcircled{0} \quad\quad 11 \dashrightarrow \textcircled{0}$$

The same method works for subtraction problems:

$$65{,}717 \dashrightarrow 14 \dashrightarrow 3$$
$$- \; 38{,}491 \dashrightarrow -9 \dashrightarrow 2$$
$$27{,}226 \dashrightarrow \textcircled{1} \quad\quad \textcircled{1}$$

It even works with multiplication:

$$853 \dashrightarrow 6$$
$$\times \quad\; 762 \dashrightarrow \times \; 3$$
$$649{,}986 \quad\quad 18$$
$$\downarrow \quad\quad\quad \downarrow$$
$$-4 \quad\quad\; \textcircled{7}$$
$$\downarrow$$
$$\textcircled{7}$$

If the numbers disagree, you made a mistake somewhere. But if they match, it's still possible for a mistake to exist. On average, this method will detect an error 10 times out of 11. Thus a mistake has a 1 in 11 chance of sneaking past the 11s check, a 1 in 9 chance of sneaking past the 9s check, and only a 1 in 99 chance of being undetected if both checks are used. For more information about this and other fascinating mathemagical topics, I would encourage you to read any of Martin Gardner's recreational math books.

You are now ready for the ultimate pencil-and-paper multiplication problem in the book, a 10-by-10! This has no practical value whatsoever except for showing off! (And personally I think multiplying 5-digit numbers is impressive enough since the answer will be beyond the capacity of most calculators.) We present it here just to prove that it can be done. The criss-crosses follow the same basic pattern as that in the 5-by-5 problem. There will be 19 computation steps and at the 10th step there will be 10 criss-crosses! Here you go:

$$2,766,829,451$$
$$\times\ 4,425,575,216$$

Here's how we did it:

STEP 1. $6 \times 1 = \underline{6}$.

STEP 2. $(6 \times 5) + (1 \times 1) = 3\underline{1}$.

STEP 3. $3 + (6 \times 4) + (2 \times 1) + (1 \times 5) = 3\underline{4}$.

STEP 4. $3 + (6 \times 9) + (5 \times 1) + (1 \times 4) + (2 \times 5) = 7\underline{6}$.

STEP 5. $7 + (6 \times 2) + (7 \times 1) + (1 \times 9) + (5 \times 5) + (2 \times 4) = 6\underline{8}$.

STEP 6. $6 + (6 \times 8) + (5 \times 1) + (1 \times 2) + (7 \times 5) + (2 \times 9) + (5 \times 4) = 13\underline{4}$.

STEP 7. $13 + (6 \times 6) + (5 \times 1) + (1 \times 8) + (5 \times 5) + (2 \times 2) + (7 \times 4) + (9 \times 5) = 16\underline{4}$.

STEP 8. $16 + (6 \times 6) + (2 \times 1) + (1 \times 6) + (5 \times 5) + (2 \times 8) + (5 \times 4) + (5 \times 2) + (7 \times 9) = 19\underline{4}$.

STEP 9. $19 + (6 \times 7) + (4 \times 1) + (1 \times 6) + (2 \times 5) + (2 \times 6) + (5 \times 4) + (5 \times 8) + (5 \times 9) + (7 \times 2) = 21\underline{2}$.

STEP 10. $21 + (6 \times 2) + (4 \times 1) + (1 \times 7) + (4 \times 5) + (2 \times 6) + (2 \times 4) + (5 \times 6) + (5 \times 9) + (7 \times 8) + (5 \times 2) = 22\underline{5}$.

STEP 11. $22 + (1 \times 2) + (4 \times 5) + (2 \times 7) + (4 \times 4) + (5 \times 6) + (2 \times 9) + (7 \times 6) + (5 \times 2) + (5 \times 8) = 21\underline{4}$.

STEP 12. $21 + (2 \times 2) + (4 \times 4) + (5 \times 7) + (4 \times 9) + (7 \times 6) + (2 \times 2) + (5 \times 6) + (5 \times 8) = 22\underline{8}$.

STEP 13. $22 + (5 \times 2) + (4 \times 9) + (7 \times 7) + (4 \times 2) + (5 \times 6) + (2 \times 8) + (5 \times 6) = 20\underline{1}$.

STEP 14. $20 + (7 \times 2) + (4 \times 2) + (5 \times 7) + (4 \times 8) + (5 \times 6) + (2 \times 6) = 15\underline{1}$.

STEP 15. $15 + (5 \times 2) + (4 \times 8) + (7 \times 5) + (4 \times 6) + (2 \times 6) = 12\underline{8}$.

STEP 16. $12 + (5 \times 2) + (4 \times 6) + (2 \times 7) + (4 \times 6) = 8\underline{4}$.

STEP 17. $8 + (2 \times 2) + (4 \times 6) + (4 \times 7) = 6\underline{4}.$

STEP 18. $6 + (4 \times 2) + (4 \times 7) = 4\underline{2}.$

STEP 19. $4 + (4 \times 2) = \underline{12}.$

If you were able to negotiate this extremely difficult problem successfully the first time through, you are on the verge of graduating from apprentice to master mathemagician!

$$\begin{array}{r} 2{,}766{,}829{,}451 \\ \times \qquad 4{,}425{,}575{,}216 \\ \hline 12{,}244{,}811{,}845{,}244{,}486{,}416 \end{array}$$

PENCIL-AND-PAPER MATHEMATICS EXERCISES

Columns of Numbers Exercises

Sum the following column of numbers, checking your answer by reading the column from bottom to top, and then by the mod sum method. If the two mod sums do not match, recheck your addition:

①
$$\begin{array}{r} 672 \\ 1367 \\ 107 \\ 7845 \\ 358 \\ 210 \\ + \quad 916 \\ \hline \end{array}$$

Sum this column of dollars and cents:

②
$$\begin{array}{r} \$ \ 21.56 \\ 19.38 \\ 211.02 \\ 9.16 \\ 26.17 \\ + \qquad 1.43 \\ \hline \end{array}$$

Subtracting-on-Paper Exercises

Subtract the following numbers, using mod sums to check your answer and then by adding the bottom two numbers to get the top number:

①
75,423
− 56,298

②
876,452
− 593,876

③
3,249,202
− 2,903,445

④
45,394,358
− 36,472,659

Square Root Exercises

For the following numbers, compute the exact square root using the doubling and dividing technique.

① $\sqrt{15}$ ② $\sqrt{502}$ ③ $\sqrt{439.2}$ ④ $\sqrt{361}$

Pencil-and-Paper Multiplication Exercises

To wrap up this exercise session, use the criss-cross method to compute exact multiplication problems of any size. Place the answer below the problems on one line, from right to left.

①
54
× 37

②
273
× 217

③
725
× 609

④
3309
× 2868

⑤
52,819
× 47,820

⑥
3,923,759
× 2,674,093

7

A Memorable Chapter

One of the most common questions I am asked pertains to my memory. I don't have an extraordinary memory. Rather, I apply a memory system that can be learned by anyone and is described in this chapter. In fact, experiments have shown that almost anyone of average intelligence can be taught to memorize numbers as well as and sometimes better than lightning calculators.

In this chapter we'll teach you a way to improve your ability to memorize numbers. Not only does this have obvious practical benefits, such as remembering dates or recalling phone numbers, it allows the mathemagician to solve much larger problems mentally. In the next chapter we'll show you how to apply the techniques of this chapter to multiply 5-digit numbers in your head!

USING MNEMONICS

The method presented in this chapter is an example of a mnemonic—that is, a tool to improve memory encoding and retrieval. (Mnemonic is pronounced neMONic, rhyming with demonic.) Mnemonics works by converting incomprehensible data (such as digit sequences) to something more meaningful. For example, take just a moment to memorize the sentence below:

"My turtle Pancho will, my love, pick up my new mover, Ginger."

Read it over several times. Look away from the page and say it to yourself over and over until you don't have to check back to the page, visualizing the turtle, Pancho, picking up your new mover, Ginger, as you do so. Got it?

A Piece of Pi for Alexander Craig Aitken

Perhaps one of the most impressive feats of mental calculation was performed by a professor of mathematics at the University of Edinburgh, Alexander Craig Aitken (1895–1967), who not only learned the value of π to 1,000 places but, when asked to demonstrate his amazing memory during a lecture, promptly rattled off the first 250 digits of π. He was then challenged to skip ahead and begin at the 551st digit and continue for another 150 places. This he did successfully without a single error.

How did he do it? Aitken explained to his audience that "the secret, to my mind, is relaxation, the complete antithesis of concentration as usually understood." Aitken's technique was more auditory than visual. He arranged the numbers into chunks of 50 digits and memorized them in a sort of rhythm. With undaunted confidence he explained: "It would have been a reprehensibly useless feat had it not been so easy."

That Aitken could memorize π to a thousand places does not qualify him as a lightning calculator. That he could easily multiply 5-digit numbers against each other does. A mathematician named Thomas O'Beirne recalled Aitken at a desk calculator demonstration. "The salesman," O'Beirne wrote, "said something like 'We'll now multiply 23,586 by 71,283.' Aitken said right off, 'And get . . .' (whatever it was). The salesman was too intent on selling even to notice, but his manager, who was watching, did. When he saw Aitken was right, he nearly threw a fit (and so did I)."

Ironically, Aitken noted that when he bought a desk calculator for himself his own mental skills quickly deteriorated. Anticipating what the future might hold, he lamented: "Mental calculators may, like the Tasmanian or the Maori, be doomed to extinction. Therefore you may be able to feel an almost anthropological interest in surveying a curious specimen, and some of my auditors here may be able to say in the year A.D. 2000, 'Yes, I knew one such.'" Fortunately, history has proved him wrong!

Congratulations! You have just memorized the first 24 digits of the mathematical expression π (pi). Recall that π is the ratio of the circumference of a circle to its diameter, usually memorized in school as 3.14. In fact, π is an irrational number (one whose digits continue indefinitely with no repetition or pattern), and computers have been used to calculate π to billions of places.

THE PHONETIC CODE

I'm sure you're wondering how "My turtle Pancho will, my love, pick up my new mover, Ginger" translates into 24 places of π.

Well, to crack this and other codes, you first need to memorize the phonetic code below in which each number between 0 and 9 is assigned a consonant sound.

1 is the "t" or "d" sound.

2 is the "n" sound.

3 is the "m" sound.

4 is the "r" sound.

5 is the "l" sound.

6 is the "j," "ch," or "sh" sound.

7 is the "k" or hard "g" sound.

8 is the "f" or "v" sound.

9 is the "p" or "b" sound.

0 is the "z" or "s" sound.

Memorizing this code isn't as hard as it looks. For one thing, notice that in cases where more than one letter is associated with a number, they have similar pronunciations. For example, the "k" sound (as it appears in words like "kite" or "cat") is similar to the hard "g" sound (as it appears in such words as "goat"). You can also rely on the following tricks to help you memorize it:

1   a typewritten "t" or "d" has just 1 downstroke.

2   a typewritten "n" has 2 downstrokes.

3   a typewritten "m" has 3 downstrokes.

4   the number 4 ends in the letter "r."

5   shape your hand with 4 fingers up and the thumb at a 90 degree angle—that's 5 fingers in an "L" shape.

6   a "J" looks like a backward 6.

7   a "K" can be drawn by laying two 7s back to back (you might call it a "special K"!).

8   a lower-case "f" written in script looks like an 8.

9   looks like a backward "p" or an upside down "b."

0   the word "zero" begins with the letter "z."

Practice remembering this list. In about 10 minutes you should have all the 1-digit numbers associated with consonant sounds. Next, you can convert numbers into words by placing vowel sounds around or between the consonant sounds. For instance, the number 32 can become any of the following words: *man, men, mine, mane, moon, many, money, menu, amen, omen, amino, mini, minnie,* and so on. Notice that the word *minnie* is acceptable since the n "sound" is only used once.

The following words could *not* represent the number 32 because they use other consonant sounds: *mourn, melon, mint.* These words would be represented by the numbers 342, 352, and 321, respectively. The consonant sounds of "h," "w," and "y" can be added freely since they don't appear on the list. Hence, 32 can also become *human, woman, yeoman, mahoney,* and so forth.

The following list gives you a good idea of the types of words you can create using this phonetic code. Don't feel obligated to memorize it—use it as inspiration to explore the possibilities on your own.

Number–Word List

1. tie	26. notch	51. light	76. cage
2. Noah	27. neck	52. lion	77. cake
3. Ma	28. knife	53. lamb	78. cave
4. ear	29. knob	54. lure	79. cap
5. law	30. mouse	55. lily	80. face
6. shoe	31. mat	56. leash	81. fight
7. cow	32. moon	57. log	82. phone
8. ivy	33. mummy	58. leaf	83. foam
9. bee	34. mower	59. lip	84. fire
10. dice	35. mule	60. cheese	85. file
11. tot	36. match	61. sheet	86. fish
12. tin	37. mug	62. chain	87. fog
13. tomb	38. movie	63. chum	88. fife
14. tire	39. map	64. cherry	89. V.I.P.
15. towel	40. rose	65. jail	90. bus
16. dish	41. rod	66. judge	91. bat
17. duck	42. rain	67. chalk	92. bun
18. dove	43. ram	68. chef	93. bomb
19. tub	44. rear	69. ship	94. bear
20. nose	45. roll	70. kiss	95. bell
21. nut	46. roach	71. kite	96. beach
22. nun	47. rock	72. coin	97. book
23. name	48. roof	73. comb	98. puff
24. Nero	49. rope	74. car	99. puppy
25. nail	50. lace	75. coal	100. daisies

For practice, translate the following numbers into words, then check the correct translation below. When translating numbers into words there are a variety of words that can be formed:

42:
74:
67:
86:
93:
10:
55:
826:
951:
620:
367:

As an exercise, translate the following words back to numbers:

dog
oven
cart
fossil
banana
bask
pencil
brown car
table tree
game of trouble
multiplication
Tony Marloshkovips

Here are some words that I came up with:

42: rain, rhino, Reno, ruin, urn
74: car, cry, guru, carry
67: jug, shock, chalk, joke, shake, hijack

86: fish, fudge
93: bum, bomb, beam, palm, pam
10: toss, dice, toes, dizzy, oats, hats
55: lily, lola, lillie, hallelujah!
826: finch, finish, vanish
951: pilot, plot, belt, bolt, bullet
620: jeans, chains, genius
367: magic!

dog: 17
oven: 82
cart: 741
fossil: 805
banana: 922
bask: 907
pencil: 9205
brown car: 94274
table tree: 19514
game of trouble: 7381495
multiplication: 35195762
Tony Marloshkovips: 1234567890

Although a number can usually be converted into many words, a word can only be translated into a single number. This is an important property for our applications as it enables us to recall specific numbers.

Using this system you can translate any number or series of numbers (e.g., phone numbers, Social Security numbers, driver's license numbers) into a word or a sentence. Here's how the code works to translate the first 24 digits of π:

3   1415   926   5   3   58   97   9   3   2   384   6264

"My turtle Pancho will, my love, pick up my new mover, Ginger."

Remember that, with this phonetic code, "g" is a hard sound, as in "grass," so a soft "g" (as in "Ginger") sounds like "j" and is represented by a 6. Also, the word "will" is, phonetically, just "L," representing one 5. Since this sentence can only be translated back to the 24 digits above, you have effectively memorized π to 24 digits!

There's no limit to the number of numbers this code will allow you to memorize. For example, the following two sentences, when added to "My turtle Pancho will, my love, pick up my new mover, Ginger," will allow you to memorize the first 60 digits of π:

3   38   327   950 2   8841   971

"My movie monkey plays in a favorite bucket."

And:

69 3   99   375 1   05820   97494

"Ship my puppy Michael to Sullivan's backrubber."

You can really feel proud of yourself once these sentences roll trippingly off your tongue, and you're able to translate them back quickly into numbers. But you've got a ways to go for the world record. Hideaki Tomoyori of Japan recited π to 40,000 places, by memory, in 17 hours and 21 minutes in 1987.

HOW MNEMONICS MAKES MENTAL CALCULATION EASIER

Aside from improving your ability to memorize long sequences of numbers, mnemonics can be used to store partial results in the middle of a difficult mental calculation. For example, here's how you can use mnemonics to help you square a 3-digit number:

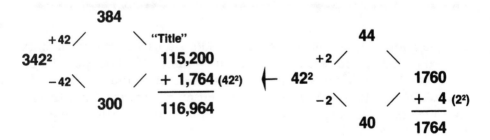

Recall that to square 342 you first multiply $300 \times 384 = 115,200$, then add 42^2. But by the time you've computed $42^2 = 1764$, you may

Oxymorons and Idiot Savants

Among the more popular terms in use in our culture today is "oxymoron," or a combination of contradictory words, applied humorously ("jumbo shrimp"), sarcastically ("military intelligence"), and sometimes curiously, as in the strange case of *idiot savant*. This term, most commonly used throughout the twentieth century to describe a mentally handicapped individual with a gift for a particular skill, comes from the time when "idiot" was an actual designation on an I.Q. test (quite low). It was coupled with "savant," the French word for "wise," and literally translates as the "wise idiot," a true oxymoron.

Many idiot savants have a remarkable talent for lightning calculations, as well as for other things, including art, music, and memory. There is the case, for example, of the Minnesota woman with an I.Q. of a two-year-old who could play the piano brilliantly without being able to read a single note of music. Or those who can memorize entire telephone directories, railway timetables, sports scores, and biographical information. There is the mentally retarded man who memorized the population figures for every town in America, the number of rooms at over 2,000 hotels, and information on over 3,000 mountains and 2,000 inventions. Among the most common skills is calendar calculating in which the idiot savant can give past and future days and dates in almost any combination of challenges, such as the years on which September 14 falls on a Wednesday. How are these individuals able to perform such mental feats? It is clear that many of the calculating skills, such as those described in this book, can be developed by anyone. But there are plenty that cannot be duplicated. One theory is that the retardation of most other skills makes available more brain power to concentrate on a single problem that can be focused on intensely and for great lengths of time. "Normal" stimulation is lacking so the idiot savant makes up for it by intense concentration on specific problems.

have forgotten the earlier number, 115,200. Here's where our memory system comes to the rescue. To store the number 115,200, put 200 on your hand by raising 2 fingers, and convert 115 into a word like "title." Repeat the word "title" to yourself once or twice. That's easier to remember than 115,200, especially after you start calculating 42^2. Once you've arrived at 1764, you can add that to "title 2," or 115,200, for a total of 116,964.

Here's another:

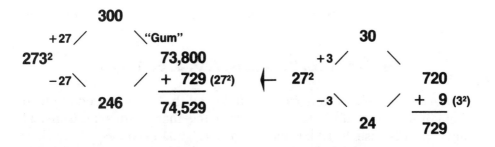

After multiplying 300 × 246 = 73,800, convert 73 into "gum" and hold 800 on your hand by raising 8 fingers or using the rule of thumb. Once you've computed $27^2 = 729$, just add that to "gum 8," or 73,800, for a total of 74,529. This may seem a bit cumbersome at first, but with practice the conversion from numbers to words and back to numbers becomes almost second nature.

You have seen how easily 2-digit numbers can be translated into simple words. Not all 3-digit numbers can be translated so easily, but if you're at a loss for a simple word to act as a mnemonic, an unusual word or a nonsense word will do. For example, if no simple word for 286 or 638 comes quickly to mind, use an old word like "knavish" or nonsense words like "jam off." Even these unusual words should be easier to recall than 286 or 638 after a long calculation. For some of the huge problems in the next chapter, these memory tricks will be indispensable.

MEMORY MAGIC

Once you've mastered the phonetic code, a whole world of memory-based magic tricks will open up to you. Practicing these tricks will also reinforce what you have just learned.

Call Out a Number

Without using mnemonics, the average human memory (including mine) can only hold on to 7 or 8 digits. But once you've mastered the ability to change numbers into words, you can expand your memory capacity considerably. Have someone slowly call out sixteen digits while someone else writes them down on a blackboard or a piece of paper. Once they're written down, you repeat them back in the exact order they were given without looking at the board or the piece of paper!

At a recent lecture demonstration, I was given the following series of numbers:

9, 7, 1, 0, 6, 8, 4, 3, 9, 2, 0, 3, 5, 5, 4, 0

As the numbers were called out, I used the phonetic code to turn them into words and then tied them together with a nonsensical story. In this case, 9710 becomes "pockets," 684 becomes "shaver," 39 becomes "map," 20 becomes "nose," 35 becomes "mule," and 540 becomes "liars."

As the numbers were being given to me I immediately began forming words and creating a silly story to help me remember them. I pictured reaching into my pockets to pull out a shaver on which there was a map with a big nose on it. On the map I saw a bunch of mules that were all liars. This story may sound bizarre, but the more ridiculous the story, the easier it is to remember—and besides, it's more fun.

8

The Tough Stuff Made Easy: Advanced Multiplication

At this point in the book—if you've gone through it chapter by chapter—you have learned to do mental addition, subtraction, multiplication, and division, as well as the art of guesstimation, pencil-and-paper mathemagics, an the phonetic code for number mnemonics. This chapter is for serious, die-hard mathemagicians who want to stretch their minds to the limits of mental calculation. The problems in this chapter range from 4-digit squares to the largest problem I perform publicly—the multiplication of two different 5-digit numbers.

In order to do these problems, it is particularly important that you know the contents of Chapter 3 quite well, and are comfortable and reasonably fast using the phonetic code. And even though, if you glance ahead in this chapter, the problems seem overwhelming, let me restate the two fundamental premises of this book: (1) All mental calculation skills can be learned by almost anyone; (2) The key is the simplification of all problems into easier problems that can be done quickly. There is no problem in this chapter, or that you will encounter anywhere (comparable to these) that cannot be mastered and learned through the simplification techniques you've learned in previous chapters. Because we are assuming that you've mastered the techniques needed for this chapter, we will be teaching primarily by illustration, rather than walking you through the problems word for word. As an aid, however, we remind you that many of the simpler problems embedded within these larger problems are ones you have already encountered in previous chapters.

We begin with 4-digit squares. Good luck!

4-DIGIT SQUARES

As a preliminary skill for mastering 4-digit squares, you need to be able to do 4-by-1 problems. We break these down into two 2-by-1 problems, as in the examples below:

$$
\begin{array}{r}
\textbf{4,867 (4,800 + 67)} \\
\times \qquad \textbf{9} \\
\hline
\end{array}
$$

$$
\begin{array}{rr}
\textbf{9} \times \textbf{4,800} = & \textbf{43,200} \\
\textbf{9} \times \textbf{67} = + & \textbf{603} \\
\hline
& \textbf{43,803}
\end{array}
$$

$$
\begin{array}{r}
\textbf{2,781 (2,700 + 81)} \\
\times \qquad \textbf{4} \\
\hline
\end{array}
$$

$$
\begin{array}{rr}
\textbf{4} \times \textbf{2,700} = & \textbf{10,800} \\
\textbf{4} \times \textbf{81} = + & \textbf{324} \\
\hline
& \textbf{11,124}
\end{array}
$$

$$
\begin{array}{r}
\textbf{6,718} \\
\times \qquad \textbf{8} \\
\hline
\end{array}
\qquad
\begin{array}{r}
\textbf{4,269} \\
\times \qquad \textbf{5} \\
\hline
\end{array}
$$

$$
\begin{array}{rr}
\textbf{8} \times \textbf{6,700} = & \textbf{53,600} \\
\textbf{8} \times \textbf{18} = + & \textbf{144} \\
\hline
& \textbf{53,744}
\end{array}
\qquad
\begin{array}{rr}
\textbf{5} \times \textbf{4,200} = & \textbf{21,000} \\
\textbf{5} \times \textbf{69} = + & \textbf{345} \\
\hline
& \textbf{21,345}
\end{array}
$$

Once you've mastered 4-by-1s you're ready to tackle 4-digit squares. Let's square 4267. Using the same method we used with 2-digit and 3-digit squares, we'll square the number 4267 by rounding down by 267 to 4000, and up by 267 to 4534, multiplying 4534 × 4000 (a 4-by-1), and then adding the square of the number you went up and down by, or 267^2, as illustrated below:

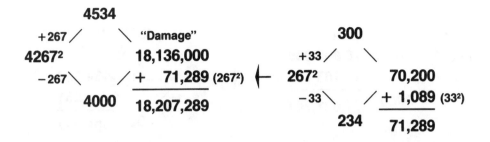

Now, obviously there is a lot going on in this problem. I realize it's one thing to say "just add the square of 267" and quite another to actually do it and remember what it was you were supposed to add it to. First of all, once you multiply 4534 × 4 to get 18,136, you can actually say the first part of the answer out loud: "18 million . . ." You can do so because the original number is always rounded to the nearest thousand. Thus the largest 3-digit number you will ever square in the next step is 500. The square of 500 is 250,000, so as soon as you see that the hundred thousands digits (in this case, 136) are less than 750, you know that the millions digit will not change.

Once you've said "18 million . . ." you need to hold on to 136,000 before you square 267. Here's where our mnemonics from the last chapter comes to the rescue! Using the phonetic code, you can convert 136 to "Damage" (1 = d, 3 = m, 6 = j). Now you can work on the next part of the problem and just remember "Damage" (and that there are 3 zeros following—this will always be the case). If at any time in the computations you forget what the original problem is you can either glance at the blackboard or, if it isn't written down, ask the audience to repeat the problem (which gives the illusion you are starting the problem over from scratch when in actual fact you have already done some of the calculations!).

You now do the 3-digit square just as you learned to before, to get 71,289. Sometimes I have trouble remembering the hundreds digits of my answer (2, in this case). I cure this by raising two of my fingers. If you forget the last two digits (89), you can go back to the original number (4267), square its last two digits ($67^2 = 4489$), and take the last two digits of that.

To compute the final answer, you now add 71,289 to "Damage" (which translates back to 136,000) resulting in 207,289, which you may now say aloud since you've already given the millions digits.

Let's do one more 4-digit square—8431^2:

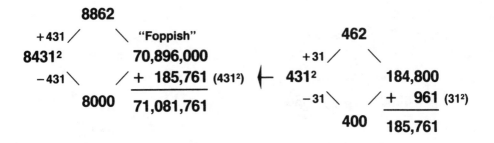

Without repeating all the detailed steps as we did in the last problem, I will point out some of the highlights of this problem. After doing the $8 \times 8862 = 70,896$, note that 896 is above 750, so a carryover is possible. In fact, since 432^2 is greater than $400^2 = 160,000$, there definitely will be a carryover when you do the addition to the number 896,000. Hence we can safely say aloud "71 million . . ." at this point.

When you square 431, you get 185,761. Add the 185 to the 896 to get 1,081, and say the rest of the answer. But remember that you already anticipated the carryover, so you just say ". . . 81 thousand . . . 761." You're done!

We illustrate one more fine point with 2753^2:

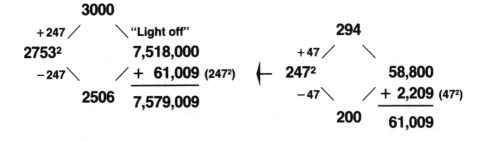

Since you are rounding up to 3000, you will be multiplying 3000 times another number in the 2000s. You could subtract $2753 - 247 = 2506$, but that's a little messy. To obtain the last three digits, double 753 to get 1506. The last three digits of this number, 506, are the last three digits of the 2000 number: 2506! This works because the two numbers being multiplied must add to twice the original number.

Then proceed in the usual manner of multiplying $3000 \times 2506 = 7,518,000$, convert 518 to the code, "Light off," and say the first part of the answer out loud as "7 million . . ." You can say this with confidence since 518 is well below 750, you know there will not be a carryover.

Thomas Fuller: Learned Men and Great Fools

It would be hard to top the physical handicap on learning experienced by Helen Keller, though the social handicap imposed on Thomas Fuller was equal in its power to stifle all but the bravest. Thomas Fuller, born in Africa in 1710, was not only illiterate; he was forced to work in the fields of Virginia as a slave and never received a single day of education. The "property" of Mrs. Elizabeth Cox, Thomas Fuller taught himself to count to 100, after which he increased his numerical powers by counting such items at hand as the grains in a bushel of wheat, the seeds in a bushel of flax, and the number of hairs in a cow's tail (2,872).

Extrapolating from mere counting, Fuller learned to compute the number of shingles a house would need to cover its roof, the number of posts and rails needed to enclose it, and other numbers relevant to materials needed in construction. His prodigious skills grew, and with them his reputation which, in his old age, brought a challenge from two Pennsylvanians to compute, in his head, numbers that would challenge the best lightning calculators. For example, they asked: "Suppose a farmer has six sows, and each sow has six female pigs the first year, and they all increase in the same proportion, to the end of eight years, how many sows will the farmer then have?" The problem can be written as $7^8 \times 6$; that is $(7 \times 7 \times 7 \times 7 \times 7 \times 7 \times 7 \times 7) \times 6$. Within ten minutes Fuller gave his response of 34,588,806, the correct answer.

Upon Fuller's death in 1790, the *Columbian Centinel* reported that "he could give the number of poles, yards, feet, inches and barley-corns in any given distance, say the diameter of the earth's orbit; and in every calculation he would produce the true answer, in less time than ninety-nine men in a hundred would take with their pens." When Fuller was asked if he regretted having never gained a traditional education he responded: "No Massa—it is best I got no learning: for many learned men be great fools."

Next, you add the square of 247. Don't forget that you can derive 247 quickly as the complement of 753 (as you learned in Chapter 3). Then proceed to the final answer as you did in the previous 4-digit problems.

Below, you will find six 4-digit square exercises to practice with. You will find the answers and computations at the back of the book.

Exercises: 4-Digit Squares

① 1234^2 ② 8639^2 ③ 5312^2 ④ 9863^2 ⑤ 3618^2 ⑥ 2971^2

3-BY-2 MULTIPLICATION

We saw in the 2-by-2 multiplication problems in Chapter 3 that there are several different ways to tackle the same problem. This variety of methods increases when you increase the number of digits in the problem. In 3-by-2 multiplication I find that it pays to take a few moments to look at the problem to determine the method of calculation that will put the least amount of strain on the brain.

Factoring Methods

The easiest 3-by-2 problems to compute are those in which the 2-digit number is factorable. For example:

$$\begin{array}{r} 637 \\ \times \;\; 56 \,(8 \times 7) \\ \hline \end{array}$$

637 × 56 = 637 × 8 × 7 = 5096 × 7 = 35,672

These are great because you don't have to add anything. You just factor 56 into 8 × 7, then do a 3-by-1 (637 × 8 = 5096), and finally a 4-by-1 (5096 × 7 = 35,672). There are no addition steps and you don't have to store any partial results.

More than half of all 2-digit numbers are factorable into numbers 11 and below, so you'll be able to use this method for many problems. Here's another example:

$$\begin{array}{r} 853 \\ \times \;\; 44 \,(11 \times 4) \\ \hline \end{array}$$

853 × 11 × 4 = 9383 × 4 = 37,532

To multiply 853 × 11, you treat 853 as 850 + 3 and proceed as follows:

$$\begin{array}{r} 850 \times 11 = \;\;\; 9350 \\ 3 \times 11 = + \;\;\; 33 \\ \hline 9383 \end{array}$$

Now multiply 9383 × 4 by treating 9383 as 9300 + 83, as follows:

$$9300 \times 4 = \quad 37{,}200$$
$$83 \times 4 = + \quad 332$$
$$\overline{ 37{,}532}$$

If the 2-digit number is not factorable, check out the 3-digit number:

$$144 \, (6 \times 6 \times 4)$$
$$\times \; 76$$

$$76 \times 144 = 76 \times 6 \times 6 \times 4 = 456 \times 6 \times 4 =$$
$$2736 \times 4 = 10{,}944$$

Notice that the sequence of the multiplication problems is a 2-by-1, a 3-by-1, and finally a 4-by-1. Since these are all problems you can now do with considerable ease, this type of 3-by-2 problem should be no problem at all.

Here's another example where the 2-digit number is not factorable but the 3-digit number is:

$$462 \, (11 \times 7 \times 6)$$
$$\times \; 53$$

$$53 \times 11 \times 7 \times 6 = 583 \times 7 \times 6 = 4081 \times 6 = 24{,}486$$

Here the sequence is a 2-by-2, a 3-by-1, and a 4-by-1, though when the 3-digit number is factorable by 11, you can use the eleven's method for a very simple 2-by-2 ($53 \times 11 = 583$). In this case it pays to be able to recognize when a number is divisible by 11 (as discussed in Chapter 4).

If the 2-digit number is not factorable, and the 3-digit number is factorable into only a 2-by-1, the problem can still be easily handled by a 2-by-2, followed by a 4-by-1:

$$423 \, (47 \times 9)$$
$$\times \; 83$$

$$83 \times 47 \times 9 = 3901 \times 9 = 35{,}109$$

Here you need to see that 423 is divisible by 9, setting up the problem as 83 × 47 × 9. The 2-by-2 is not easy, but by treating 83 as 80 + 3, you get:

$$
\begin{array}{r}
\textbf{83 (80 + 3)} \\
\times \quad \textbf{47} \\
\hline
\textbf{80 × 47 = } \quad \textbf{3760} \\
\textbf{3 × 47 = + 141} \\
\hline
\textbf{3901}
\end{array}
$$

Then do the 4-by-1 problem of 3901 × 9 for your final answer of 35,109.

The Addition Method

If the 2-digit number and the 3-digit number cannot be conveniently factored in the 3-by-2 problem you're doing, you will usually resort to the addition method:

$$
\begin{array}{r}
\textbf{721 (720 + 1)} \\
\times \quad \textbf{37} \\
\hline
\textbf{720 × 37 = } \quad \textbf{26,640} \text{ (treating 72 as 9 × 8)} \\
\textbf{1 × 37 = + 37} \\
\hline
\textbf{26,677}
\end{array}
$$

The method requires you to add a 2-by-2 (times 10) to a 2-by-1. These problems tend to be more difficult than problems that can be factored since you have to perform a 2-by-1 while holding on to a 5-digit number, and then add the results together. In fact, it is probably easier to solve this problem by factoring 721 into 103 × 7, then computing 37 × 103 × 7 = 3811 × 7 = 26,677.

As another example:

$$
\begin{array}{r}
\textbf{732 (730 + 2)} \\
\times \quad \textbf{57} \\
\hline
\textbf{730 × 57 = } \quad \textbf{41,610} \text{ (treating 73 as 70 + 3)} \\
\textbf{2 × 57 = + 114} \\
\hline
\textbf{41,724}
\end{array}
$$

Though you will usually break up the 3-digit number when using the addition method, it occasionally pays to break up the 2-digit number instead, particularly when the last digit of the 2-digit number is a 1 or a 2, as in the following example:

$$\begin{array}{r} 386 \\ \times \quad 51 \text{ (50 + 1)} \end{array}$$

$$\begin{array}{rr} 50 \times 386 = & 19,300 \\ 1 \times 386 = + & 386 \\ \hline & 19,686 \end{array}$$

This reduces the 3-by-2 to a 3-by-1, made especially easy since the second multiplication problem involves a 1. Note too that we were aided in multiplying a 5 by an even number, which produces an extra 0 in the answer, so there is only one digit of overlap in the addition problem.

Another example of multiplying a 5 by an even number is illustrated in the following problem:

$$\begin{array}{r} 835 \\ \times \quad 62 \text{ (60 + 2)} \end{array}$$

$$\begin{array}{rr} 60 \times 835 = & 51,000 \\ 2 \times 835 = + & 1,670 \\ \hline & 52,670 \end{array}$$

When you multiply the 6 in 60 by the 5 in 835, it generates an extra 0 in the answer, making the addition problem especially easy.

The Subtraction Method

As with 2-by-2s, it is sometimes more convenient to solve a 3-by-2 with subtraction instead of addition, as in the following problems:

$$\begin{array}{r} 629 \text{ (630 − 1)} \\ \times \quad 38 \end{array}$$

$$\begin{array}{rr} 630 \times 38 = & 23,940 \text{ (63 = 9 × 7)} \\ -1 \times 38 = - & 38 \\ \hline & 23,902 \end{array}$$

$$758\ (760 - 2)$$
$$\times\qquad 43$$

$$760 \times 43 = \quad 32{,}680 \ (43 = 40 + 3)$$
$$-2 \times 43 = -\qquad 86$$
$$\overline{32{,}594}$$

By contrast, you can compare the last subtraction method with the addition method, below, for this same problem:

$$758\ (750 + 8)$$
$$\times\qquad 43$$

$$750 \times 43 = \quad 32{,}250 \ (75 = 5 \times 5 \times 3)$$
$$8 \times 43 = +\quad 344$$
$$\overline{32{,}594}$$

My preference for tackling this problem would be to use the subtraction method because I always try to leave myself with the easiest possible addition or subtraction problem at the end. In this case, I'd rather subtract 86 than add 344, even though the 2-by-2 multiplication problem in the subtraction method above is slightly harder than the one in the addition method.

The subtraction method can also be used for 3-digit numbers below a multiple of 100 or close to 1000, as in the next two examples:

$$293\ (300 - 7)$$
$$\times\qquad 87$$

$$300 \times 87 = \quad 26{,}100$$
$$-7 \times 87 = -\quad 609$$
$$\overline{25{,}491}$$

$$988\ (1000 - 12)$$
$$\times\qquad 68$$

$$1000 \times 68 = \quad 68{,}000$$
$$-12 \times 68 = -\qquad 816 \ (12 = 6 \times 2)$$
$$\overline{67{,}184}$$

The last three digits of the answers were obtained by taking the complements of 609 − 100 = 509 and 816 respectively.

Finally, in the following illustration we break up the 2-digit number using the subtraction method. Notice how we subtract 736 by subtracting 1000 and adding back the complement:

$$
\begin{array}{r}
\mathbf{736} \\
\times \quad \mathbf{59}\ \mathbf{(60-1)} \\
\end{array}
$$

60 × 736 =	**44,160**	**44,160**
−1 × 736 =	**− 736**	**− 1,000**
	43,424	**43,160**
		+ 264 (complement of 736)
		43,424

3-by-2 Exercises Using Factoring, Addition, and Subtraction Methods

Solve the 3-by-2 problems below, using the factoring, addition, or subtraction method. Factoring, when possible, is usually easier. Our solutions appear in the back of the book.

①
$$
\begin{array}{r}
858 \\
\times\ 15 \\
\end{array}
$$

②
$$
\begin{array}{r}
796 \\
\times\ 19 \\
\end{array}
$$

③
$$
\begin{array}{r}
148 \\
\times\ 62 \\
\end{array}
$$

④
$$
\begin{array}{r}
773 \\
\times\ 42 \\
\end{array}
$$

⑤
$$
\begin{array}{r}
906 \\
\times\ 46 \\
\end{array}
$$

⑥
$$
\begin{array}{r}
952 \\
\times\ 26 \\
\end{array}
$$

⑦
$$
\begin{array}{r}
411 \\
\times\ 93 \\
\end{array}
$$

⑧
$$
\begin{array}{r}
967 \\
\times\ 51 \\
\end{array}
$$

⑨
$$
\begin{array}{r}
484 \\
\times\ 75 \\
\end{array}
$$

⑩
$$
\begin{array}{r}
126 \\
\times\ 87 \\
\end{array}
$$

⑪
$$
\begin{array}{r}
157 \\
\times\ 33 \\
\end{array}
$$

⑫
$$
\begin{array}{r}
616 \\
\times\ 37 \\
\end{array}
$$

⑬
$$
\begin{array}{r}
841 \\
\times\ 72 \\
\end{array}
$$

⑭
$$
\begin{array}{r}
361 \\
\times\ 41 \\
\end{array}
$$

⑮
$$
\begin{array}{r}
218 \\
\times\ 68 \\
\end{array}
$$

⑯
$$
\begin{array}{r}
538 \\
\times\ 53 \\
\end{array}
$$

⑰
817
× 61

⑱
668
× 63

⑲
499
× 25

⑳
144
× 56

㉑
281
× 44

㉒
988
× 22

㉓
383
× 49

Jedediah Buxton: 5,116 Pints of Genius

We are strangely compelled by stories of ordinary people doing extraordinary feats so far beyond expectation that they invoke awe and wonder. Such is the case of Jedediah Buxton (1707–1772), an illiterate English farmer from the tiny town of Elmton, who, while working in the field one day, was approached by the journalist George Saxe who had heard of his ability to do mental calculations. Saxe proposed to him the following: "In a body whose 3 sides are 23,145,789 yards, 5,642,732 yards, and 54,965 yards, how many cubic inches are there?" Five hours later Saxe returned to hear Buxton's immediate recital of a 28-digit answer that was off by only 1 digit. (Buxton offered to repeat it backwards, as well.) Describing Buxton as "this surprising genius now cloathed in rags and labouring hard with his spade for the support of himself and a large family," Saxe reported that he could also survey land almost as accurately by pacing off the perimeter as it could be measured by chain.

Buxton attained some prominence when the popular *Gentleman's Magazine* sent a writer to test his calculating skills for an article. Buxton was given a simple 3-by-3 multiplication problem (the area of a field 423 by 383 yards), which he answered in under two minutes (162,009 yards). They then asked him to compute the number of barleycorns needed to reach 8 miles. He calculated 1,520,640, at 3 to the inch. Buxton even correctly figured the number of revolutions a coachwheel 6 yards in circumference would make in going the 204 miles from York to London—59,840 times.

Though Buxton never got rich from his talents, he managed to hustle free beer from challengers of his skill, logging a mental record of 5,116 pints since age 12, an average of five to six ounces per day. The drink did not seem to slow his mental powers, however, as it was reported that he often partook of the ale while computing monstrous problems!

The following 3-by-2s will appear in the 5-digit squares and the 5-by-5 multiplication problems that follow.

㉔
589
× 87

㉕
286
× 64

㉖
853
× 32

㉗
878
× 24

㉘
423
× 45

㉙
154
× 19

㉚
834
× 34

㉛
545
× 27

㉜
653
× 69

㉝
216
× 78

㉞
822
× 95

5-Digit Squares

Mastering 3-by-2 multiplication takes a fair amount of practice, but once you've got that done you can slide right into doing 5-digit squares because they simplify into a 3-by-2 problem plus a 2-digit square and a 3-digit square. Sound too easy? Watch:

To square the following number:

$$46,792^2$$

Treat it as:

$$46,000 + 792$$
$$\times\ 46,000 + 792$$

Using the distributive law we can break this down to:

1.   **2.**   **3.**
$$46,000 \times 46,000 + 2(46,000)(792) + (792)(792)$$

This can be simplifed further as:

$$46^2 \times 1\ \text{million} + (46)(792)(2000) + (792)^2$$

But I do not do them in this order. In fact, I start in the middle, because the 3-by-2 problems are harder than the 2-digit and 3-digit squares. So in keeping with the principle of getting the hard stuff out of the way first, I do 792 × 46 × 2 and attach 3 zeros on the end, as follows:

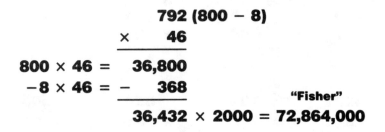

Using the subtraction method, as shown above, compute 792 × 46 = 36,432, then double that number to get 72,864. Using the phonetic code from Chapter 7 on the number 864 allows you to store this number as "72 Fisher."

The next step is to do 46^2 × 1 million, which is **2,116,000,000.**

At this point, you can say, **"2 billion . . ."**

Recalling the 72 of "72 Fisher," you add 116 million to 72 million = 188 million. Before saying this number aloud you need to check ahead to see if there is a carryover when adding "Fisher," or 864, to 792^2. Here you don't actually calculate 792^2; rather, you see that its product will be large enough to make the 864,000 carry over. (You can guesstimate this by noting that 800^2 is 640,000, which will easily make the 864,000 carry over, thus you bump the 188 up a notch and say, **". . . 189 million . . ."**

Still holding on to the "Fisher," compute the square of 792, using the 3-digit square method from Chapter 3 (rounding up and down by 8, etc.) to get 627,264.

To make the final computation you add 627 to "Fisher," or 864, to get 1491. But since you already made the carryover, drop the 1 and say, **". . . 491 thousand 264."**

Sometimes I forget the last three digits of the answer because my mind has been so preoccupied with the larger computations. So before doing the final addition I will store the 2 of 264 on my fingers and try to remember the 64, which I can usually do because we tend to recall the most recent things heard. If this fails, I can come up with the final two digits by squaring the final two digits of the original number, 92^2, or 8464, the last two digits of which are the last two digits I'm looking for: **64.**

I know this is quite a mouthful. To reiterate the entire problem in a single illustration, here is how I computed **46,792²:**

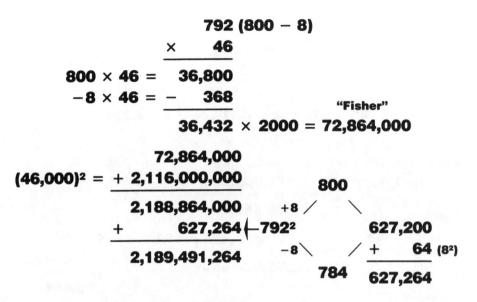

Let's look at another 5-digit square example:

$$83,522^2$$

As before, we compute, in order:

83 × 522 × 2000, 83² × 1 million, then (522)².

For the first problem, notice that 522 is a multiple of 9. In fact, $522 = 58 \times 9$. Treating 83 as 80 + 3, we get:

$$
\begin{array}{r}
\textbf{522 (58 × 9)} \\
\times\ \textbf{83} \\
\hline
\end{array}
$$

83 × 58 × 9 = 4814 × 9 = 43,326

Doubling 43,326 results in 86,652, which can be stored as "86 Julian." Since $83^2 = 6889$, we can say, **"6 billion . . ."**

Adding 889 + 86 gives us 975. Before saying 975 million we check to see if "Julian" (652,000) will carry over after squaring 522. Guesstimating 522^2 as about 270,000 (500 × 540) I see it will not carry over. Thus you can safely say: **". . . 975 million . . ."**

Finally, you square 522 in the usual way to get 272,484 and add that to "Julian" (652,000) for the rest of the answer: **". . . 924,484."**

Illustrated, this problem looks like:

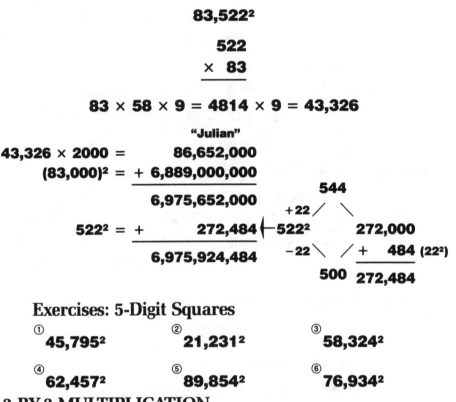

$$83{,}522^2$$

$$\begin{array}{r} 522 \\ \times\ 83 \\ \hline \end{array}$$

$$83 \times 58 \times 9 = 4814 \times 9 = 43{,}326$$

"Julian"

$$\begin{array}{rr} 43{,}326 \times 2000 = & 86{,}652{,}000 \\ (83{,}000)^2 = & +\ 6{,}889{,}000{,}000 \\ \hline & 6{,}975{,}652{,}000 \\ 522^2 = & +\qquad 272{,}484 \\ \hline & 6{,}975{,}924{,}484 \end{array}$$

$$\begin{array}{ccc} & 544 & \\ +22\nearrow & & \searrow \\ 522^2 & & 272{,}000 \\ -22\searrow & & \nearrow\ +\quad 484\ (22^2) \\ & 500 & 272{,}484 \end{array}$$

Exercises: 5-Digit Squares

① $45{,}795^2$ ② $21{,}231^2$ ③ $58{,}324^2$

④ $62{,}457^2$ ⑤ $89{,}854^2$ ⑥ $76{,}934^2$

3-BY-3 MULTIPLICATION

In building to the grand finale of 5-by-5 multiplication, 3-by-3s are the final hurdle. As with 3-by-2s, there are a number of methods you can use to exploit the numbers in the simplification process.

Factoring Method

We'll begin with the factoring method. Unfortunately, most 3-digit numbers are not factorable into 1-digit numbers, but when they are, the calculation is fairly simple.

$$\begin{array}{r} 829 \\ \times\ 288\ (9 \times 8 \times 4) \\ \hline \end{array}$$

$$829 \times 9 \times 8 \times 4 = 7461 \times 8 \times 4 = 59{,}688 \times 4 = \\ 238{,}752$$

Notice the sequence involved. You simplify the 3-by-3 (829 × 288) to a 3-by-1-by-1-by-1 through the factoring of 288 into 9 × 8 × 4. This then turns into a 4-by-1-by-1 (7461 × 8 × 4), and finally into a

5-by-1 to yield the final answer of 238,752. The beauty of this process is that there is nothing to add and nothing to store in memory. When you get to the 5-by-1, you are one step away from completion.

The 5-by-1 problem can be solved in two steps by treating 59,688 as 59,000 + 688, then adding the results of the 2-by-1 (59,000 × 4) and the 3-by-1 (658 × 4), as below:

$$\begin{array}{r} \textbf{59,688 (59,000 + 688)} \\ \times \quad \textbf{4} \\ \hline \end{array}$$

$$\begin{array}{r} \textbf{59,000} \times \textbf{4} = \textbf{236,000} \\ \textbf{688} \times \textbf{4} = + \;\textbf{2,752} \\ \hline \textbf{238,752} \end{array}$$

If both 3-digit numbers are factorable into 2-by-1s, then the 3-by-3 can be simplified to a 2-by-2-by-1-by-1, as in the following problem:

$$\begin{array}{r} \textbf{513 (57} \times \textbf{9)} \\ \times \; \textbf{246 (41} \times \textbf{6)} \\ \hline \end{array}$$

$$\begin{array}{l} \textbf{57} \times \textbf{41} \times \textbf{9} \times \textbf{6} \\ = \textbf{2337} \times \textbf{9} \times \textbf{6} \\ = \textbf{21,033} \times \textbf{6} \\ = \textbf{126,198} \end{array}$$

As usual, it is best to get the hard part of the problem over first (the 2-by-2). Once you've got this, the problem is then reduced to a 4-by-1, then to a 5-by-1.

More often than not, only one of the 3-digit numbers will be factorable, in which case it is reduced to a 3-by-2-by-1, as in the following problem:

$$\begin{array}{r} \textbf{459 (51} \times \textbf{9)} \\ \times \; \textbf{526} \\ \hline \end{array}$$

$$\begin{array}{l} \textbf{526} \times \textbf{459} \\ = \textbf{526} \times \textbf{51} \times \textbf{9} \\ = \textbf{526} \times \textbf{(50 + 1)} \times \textbf{9} \\ = \textbf{26,826} \times \textbf{9} \\ = \textbf{241,434} \end{array}$$

The next 3-by-3 is really just a 3-by-2 in disguise:

$$624$$
$$\times\ 435$$

By doubling the 435 and cutting 624 in half, we obtain the equivalent problem:

312 (52 × 6)
× 870 (87 × 10)

$$87 \times 52 \times 6 \times 10$$
$$= 87 \times (50 + 2) \times 6 \times 10$$
$$= 4524 \times 6 \times 10$$
$$= 27{,}144 \times 10$$
$$= 271{,}440$$

Close-Together Method

The next multiplication shortcut is based on the following algebraic formula:

$$(z + a)(z + b) = z^2 + za + zb + ab$$

Which we rewrite as:

$$(z + a)(z + b) = z(z + a + b) + ab$$

This formula is valid for any values of z, a, and b. We shall take advantage of this whenever the 3-digit numbers to be multiplied (z + a) and (z + b) are both near a number, z (typically with lots of zeros in it). For example, to multiply:

107
× 111

We treat this problem as (100 + 7)(100 + 11).
Using z = 100, a = 7, b = 11, our formula gives us:

$$100\ (100 + 7 + 11) + 7 \times 11$$
$$= 100 \times 118 + 77$$
$$= 11{,}877$$

I diagram the problem this way:

$$
\begin{array}{r}
\textbf{107 (7)} \\
\times \quad \textbf{111 (11)} \\
\hline
\end{array}
$$

$$
\begin{array}{rr}
\textbf{100} \times \textbf{118} = & \textbf{11,800} \\
\textbf{7} \times \textbf{11} = + & \textbf{77} \\
\hline
& \textbf{11,877}
\end{array}
$$

The numbers in parentheses denote the difference between the number and our convenient "base number" (here, z = 100). The number 118 can be obtained either by adding 107 + 11 or 111 + 7. Algebraically these sums are always equal since $(z + a) + b = (z + b) + a$.

With fewer words this time, here's another quickie:

$$
\begin{array}{r}
\textbf{109 (9)} \\
\times \quad \textbf{104 (4)} \\
\hline
\end{array}
$$

$$
\begin{array}{rr}
\textbf{100} \times \textbf{113} = & \textbf{11,300} \\
\textbf{9} \times \textbf{4} = + & \textbf{36} \\
\hline
& \textbf{11,336}
\end{array}
$$

Neat!

Let's up the ante a little with a higher base number:

$$
\begin{array}{r}
\textbf{408 (8)} \\
\times \quad \textbf{409 (9)} \\
\hline
\end{array}
$$

$$
\begin{array}{rr}
\textbf{400} \times \textbf{417} = & \textbf{166,800} \\
\textbf{8} \times \textbf{9} = + & \textbf{72} \\
\hline
& \textbf{166,872}
\end{array}
$$

Although this method is usually used for 3-digit multiplication, we can use it for 2-by-2s, as well:

$$
\begin{array}{r}
\textbf{78 (8)} \\
\times \quad \textbf{73 (3)} \\
\hline
\end{array}
$$

$$
\begin{array}{rr}
\textbf{70} \times \textbf{81} = & \textbf{5670} \\
\textbf{8} \times \textbf{3} = + & \textbf{24} \\
\hline
& \textbf{5694}
\end{array}
$$

Here, the base number is 70, which we multiply by 81 (78 + 3). Even the addition component is usually very simple.

We can also apply this method when the two numbers are both *lower* than the base number, as in the following problem where both numbers are just under 400:

$$
\begin{array}{r}
396\,(-4) \\
\times \quad 387\,(-13) \\
\hline
\end{array}
$$

$$
\begin{array}{rr}
400 \times 383 = & 153{,}200 \\
-4 \times -13 = + & 52 \\
\hline
& 153{,}252
\end{array}
$$

The number 383 can be obtained from 396 − 13, or from 387 − 4. I would use this method for 2-by-2 problems like the ones below:

$$
\begin{array}{r}
97\,(-3) \\
\times \quad 94\,(-6) \\
\hline
\end{array}
$$

$$
\begin{array}{rr}
100 \times 91 = & 9100 \\
-3 \times -6 = + & 18 \\
\hline
& 9118
\end{array}
$$

$$
\begin{array}{r}
79\,(-1) \\
\times \quad 78\,(-2) \\
\hline
\end{array}
$$

$$
\begin{array}{rr}
80 \times 77 = & 6160 \\
-1 \times -2 = + & 2 \\
\hline
& 6162
\end{array}
$$

In our next example, the base number falls *between* the two numbers:

$$
\begin{array}{r}
396\,(-4) \\
\times \quad 413\,(13) \\
\hline
\end{array}
$$

$$
\begin{array}{rr}
400 \times 409 = & 163{,}600 \\
-4 \times 13 = - & 52 \\
\hline
& 163{,}548
\end{array}
$$

The number 409 is obtained from 396 + 13, or 413 − 4. Notice that since −4 and 13 are of opposite signs we must subtract 52 here.

Let's raise the ante higher still, to where the second step requires a 2-by-2 multiplication:

$$
\begin{array}{r}
\textbf{621 (21)} \\
\times \quad \textbf{637 (37)} \\
\hline
\end{array}
$$

600 × 658 =	**394,800**	
21 × 37 = +	**777**	**(37 × 7 × 3)**

$$\underline{}$$

395,577

We note here that step 1 in the multiplication problem (600 × 658) is the product obtained by the guesstimation method we met in Chapter 5. Our method enables you to go from a reasonable guesstimate to an exact answer.

$$
\begin{array}{r}
\textbf{876 (−24)} \\
\times \quad \textbf{853 (−47)} \\
\hline
\end{array}
$$

900 × 829 =	**746,100**	
(−24) × (−47) = +	**1,128**	**(47 × 6 × 4)**

747,228

Also notice that in all these examples, the numbers we multiply in the first step have the same sum as the original numbers. For example, in the problem above, 900 + 829 = 1729 just as 876 + 853 = 1729. This is because:

$$z + [(z + a) + b] = (z + a) + (z + b)$$

Thus, to obtain the number to be multiplied by 900 (which you know will be 800 plus something), you need only look at the last two digits of 76 + 53 = 129 to determine 829.

In the next problem, adding 827 + 761 = 1588 tells us that we should multiply 800 × 788, then *subtract* 27 × 39 as follows:

$$
\begin{array}{r}
\mathbf{827\ (+27)} \\
\times \quad \mathbf{761\ (-39)} \\
\hline
\end{array}
$$

$$
\begin{array}{rl}
\mathbf{800 \times 788 =} & \mathbf{630,400} \\
\mathbf{-39 \times 27 =} & \mathbf{-\quad 1,053\ (39 \times 9 \times 3)} \\
\hline
& \mathbf{629,347}
\end{array}
$$

This method is so effective that if the 3-by-3 problem you are presented with has numbers that are not close together, you can sometimes modify the problem by dividing one and multiplying the other, both by the same number, to bring them closer together. For instance, 672×157 can be solved by:

$$
\begin{array}{rl}
\mathbf{672 \div 2 =} & \mathbf{336\ (36)} \\
\mathbf{\times\ 157 \times 2 =\ \times} & \mathbf{314\ (14)} \\
\hline
\mathbf{300 \times 350 =} & \mathbf{105,000} \\
\mathbf{36 \times 14 =\ +} & \mathbf{504\ (36 \times 7 \times 2)} \\
\hline
& \mathbf{105,504}
\end{array}
$$

When the numbers being multiplied are the same (you can't get any closer than that!) notice that with the close-together method you perform the exact same calculations you did in our traditional squaring procedure:

$$
\begin{array}{r}
\mathbf{347\ (47)} \\
\times \quad \mathbf{347\ (47)} \\
\hline
\end{array}
$$

$$
\begin{array}{rl}
\mathbf{300 \times 394 =} & \mathbf{118,200} \\
\mathbf{47^2 =\ +} & \mathbf{2,209} \\
\hline
& \mathbf{120,409}
\end{array}
$$

$$
\begin{array}{ccc}
 & \mathbf{394} & \\
\mathbf{+47}\nearrow & & \searrow \\
\mathbf{347^2} & & \mathbf{118,200} \\
\mathbf{-47}\searrow & & \nearrow\ \mathbf{+\ 2,209}\ (47^2) \\
 & \mathbf{300} & \mathbf{120,409}
\end{array}
$$

Addition Method

When none of the previous methods works, I look for an addition method possibility, particularly when the numbers of the first two digits of one of the 30-digit numbers is easy to work with. For example, in the problem below, the 64 of 641 is factorable into 8 × 8, so I would solve the problem as illustrated:

$$
\begin{array}{r}
373 \\
\times \quad\quad 641 \;(640 + 1) \\
\hline
\end{array}
$$

640 × 373 =	238,720	(373 × 8 × 8 × 10)
1 × 373 = +	373	

$$239,093$$

In a similar way, in the next problem the 42 of 427 is factorable into 7 × 6, so you can use the addition method and treat 427 as 420 + 7:

$$
\begin{array}{r}
656 \\
\times \quad\quad 427 \;(420 + 7) \\
\hline
\end{array}
$$

420 × 656 =	275,520	(656 × 7 × 6 × 10)
7 × 656 =	4,592	

$$280,112$$

Often I break the last addition problem into two steps, as follows:

	275,520
7 × 600 = +	4,200

	279,720
7 × 56 = +	392

$$280,112$$

Since addition-method problems can be very strenuous, I usually go out of my way to find a method that will produce a simple addition computation at the end. For example, the above problem could have been done using the factoring method. In fact, that is how I would choose to do it:

$$656$$
$$\times\ 427\ (61 \times 7)$$

$$656 \times 61 \times 7$$
$$= 656 \times (60 + 1) \times 7$$
$$= 40{,}016 \times 7$$
$$= 280{,}112$$

The simplest addition-method problems are those in which one number has a 0 in the middle, as below:

$$732$$
$$\times\qquad 308\ (300 + 8)$$

$$300 \times 732 =\ \ \ 219{,}600$$
$$8 \times 732 = +\ \ \ \ \ 5{,}856$$

$$225{,}456$$

These problems tend to be so much easier than other addition-method problems that it pays to see whether the 3-by-3 can be converted to a problem like this. For instance, 732×308 could have been obtained by either of the "non-zero" problems below:

$$244 \times 3 =\ \ \ \ 732$$
$$\times\ 924 \div 3 = \times\ 308$$

or

$$366 \times 2 =\ \ \ \ 732$$
$$\times\ 616 \div 2 = \times\ 308$$

Let's do one more toughie:

$$739$$
$$\times\qquad 443\ (440 + 3)$$

$$440 \times 739 =\ \ \ 325{,}160\ (739 \times 11 \times 4 \times 10)$$
$$3 \times 700 = +\ \ \ \ 2{,}100$$

$$327{,}260$$
$$3 \times 39 = +\ \ \ \ \ \ 117$$

$$327{,}377$$

Subtraction Method

The subtraction method is one that I sometimes use when one of the 3-digit numbers can be rounded up to a convenient 2-digit number with a 0 at the end, as in the next problem:

$$
\begin{array}{r}
\textbf{719 (720 } - \textbf{ 1)} \\
\times \quad \textbf{247} \\
\hline
\end{array}
$$

$$
\begin{array}{rl}
\textbf{720} \times \textbf{247} = & \textbf{177,840 } \textbf{(247} \times \textbf{9} \times \textbf{8} \times \textbf{10)} \\
-\textbf{1} \times \textbf{247} = & -\quad\ \ \textbf{247} \\
\hline
& \textbf{177,593}
\end{array}
$$

Likewise, in the following example:

$$
\begin{array}{r}
\textbf{538 (540 } - \textbf{ 2)} \\
\times \quad \textbf{346} \\
\hline
\end{array}
$$

$$
\begin{array}{rl}
\textbf{540} \times \textbf{346} = & \textbf{186,840 } \textbf{(346} \times \textbf{6} \times \textbf{9} \times \textbf{10)} \\
-\textbf{2} \times \textbf{346} = & -\quad\ \ \textbf{692} \\
\hline
& \textbf{186,148}
\end{array}
$$

WHEN-ALL-ELSE-FAILS METHOD

When all else fails, as they say, I use the following method, which is foolproof when you can find no other method to exploit the numbers. In the when-all-else-fails method, the 3-by-3 problem is broken down into three parts: a 3-by-1, a 2-by-1, and a 2-by-2. As you do each computation, you sum the totals as you go. These problems are difficult, especially if you cannot see the original number. In my presentation of 3-by-3s and 5-by-5s, I have the problems written down, but I do all my calculations mentally. Here's an example:

$$
\begin{array}{r}
\textbf{851} \\
\times \quad \textbf{527} \\
\hline
\end{array}
$$

$$
\begin{array}{rl}
\textbf{500} \times \textbf{851} = & \textbf{425,500} \\
\textbf{27} \times \textbf{800} = & +\ \ \textbf{21,600} \\
\hline
& \textbf{447,100} \\
\textbf{27} \times \textbf{51} = & +\quad\ \ \textbf{1,377 } \textbf{[27} \times \textbf{(50 + 1)]} \\
\hline
& \textbf{448,477}
\end{array}
$$

In practice, the calculation actually proceeds as shown below. Sometimes I use the phonetic code to store the thousand digits (e.g., 447 = "our rug") and put the hundreds digit (1) on my fingers:

$$
\begin{array}{r}
851 \\
\times\ 527 \\
\end{array}
$$

$$
\begin{array}{rr}
5 \times 851 = & 4{,}255 \\
8 \times 27 = & +\ \ 216 \\
\hline
\end{array}
$$

"Our rug"

$$
\begin{array}{rr}
4{,}471 \times 100 = & 447{,}100 \\
51 \ \times 27 = & +\ \ \ 1{,}377 \\
\hline
& 448{,}477 \\
\end{array}
$$

Let's do another example, but this time I'll break up the first number. (I usually break up the larger one so that the addition problem is easier.)

$$
\begin{array}{r}
923 \\
\times\ 673 \\
\end{array}
$$

$$
\begin{array}{rr}
9 \times 673 = & 6{,}057 \\
6 \times 23 = & +\ \ 138 \\
\hline
\end{array}
$$

"Shut up"

$$
\begin{array}{rr}
6{,}195 \times 100 = & 619{,}500 \\
73 \ \times 23 = & +\ \ \ 1{,}679 \\
\hline
& 621{,}179 \\
\end{array}
$$

3-by-3 Exercises

①
$$
\begin{array}{r}
644 \\
\times\ 286 \\
\end{array}
$$

②
$$
\begin{array}{r}
644 \\
\times\ 286 \\
\end{array}
$$

③
$$
\begin{array}{r}
596 \\
\times\ 167 \\
\end{array}
$$

④
$$
\begin{array}{r}
853 \\
\times\ 325 \\
\end{array}
$$

⑤
$$
\begin{array}{r}
343 \\
\times\ 226 \\
\end{array}
$$

⑥
$$
\begin{array}{r}
809 \\
\times\ 527 \\
\end{array}
$$

⑦
$$
\begin{array}{r}
942 \\
\times\ 879 \\
\end{array}
$$

⑧
$$
\begin{array}{r}
692 \\
\times\ 644 \\
\end{array}
$$

⑨
$$
\begin{array}{r}
446 \\
\times\ 176 \\
\end{array}
$$

⑩
$$
\begin{array}{r}
658 \\
\times\ 468 \\
\end{array}
$$

⑪
$$
\begin{array}{r}
273 \\
\times\ 138 \\
\end{array}
$$

⑫
$$
\begin{array}{r}
824 \\
\times\ 206 \\
\end{array}
$$

⑬
  642
× 249

⑭
  783
× 589

⑮
  871
× 926

⑯
  341
× 715

⑰
  417
× 298

⑱
  557
× 756

⑲
  976
× 878

⑳
  765
× 350

The following problems are embedded in the 5-by-5 multiplication problems in the next section:

㉑
  154
× 423

㉒
  545
× 834

㉓
  216
× 653

㉔
  393
× 822

5-BY-5 MULTIPLICATION

The highest we will go in mental multiplication is 5-by-5 problems, for two reasons. First, you could do 6-by-6s (by treating them as two 3-digit numbers each), but they take considerably longer than 5-by-5s and the point of mathemagics is *rapid* mental multiplication. Second, even if you could do 6-by-6s quickly, most calculators do not have a large enough capacity to check the answer!

To do a 5-by-5, you need to have mastered 2-by-2s, 2-by-3s, and 3-by-3s, as well as the phonetic code. By the time you've gotten this far in mathemagics, the earlier stuff has probably become almost second nature to you. Thus it is just a matter of putting it all together. As you did in the 5-digit square problems, you will use the distributive law to break down the numbers. For example:

$$27,639 \; (27,000 + 639)$$
$$\times \; 52,196 \; (52,000 + 196)$$

Based on this you can break the problem down into four easier multiplication problems, which I illustrate below in a criss-cross fashion as a 2-by-2, two 3-by-2s, and finally a 3-by-3, summing the results for a grand total. That is,

$$(27 \times 52) \text{ million}$$
$$+ \; [(27 \times 196) + (52 \times 639)] \text{ thousand}$$
$$+ \; (639 \times 196)$$

As with the 5-digit squares, I start in the middle with the 3-by-2s, starting with the harder 3-by-2:

"Mom no knife"

1. **52 × 639 = 52 × 71 × 9 = 3692 × 9 = 33,228**

Committing 33,228 to memory with the mnemonic "Mom no knife," you then turn to the second 3-by-2:

2. **27 × 196 = 27 × (200 − 4) = 5400 − 108 = 5292,**

and add it to the number that you are storing:

3.   **33,228** ("Mom no knife")
     + 5,292
       ─────────
      38,520

for a new total, which we store as:

"Movie lines" (38 million, 520 thousand)

Holding on to "movie lines," compute the 2-by-2:

4. **52 × 27 = 52 × 9 × 3 = 1,404**

At this point, you can give part of the answer. Since this 2-by-2 represents 52 × 27 *million*, 1,404 represents 1 billion, 404 million. Since 404 million will not carry over, you can safely say: **"1 billion . . ."**

5. **404 + "Movie" (38) = 442**

In this step you add 404 to "Movie" (38) = 442, at which point you can say: **". . . 442 million . . ."** You can say this because you know 442 will not carry over—you've peeked ahead at the 3-by-3 to see whether it will cause 442 to carry over to a higher number. If you found that it would carry over, you would say "443 million." But since "lines" is 520,000 and the 3-by-3 (639 × 196) will not exceed 500,000 (a rough guesstimate of 600 × 200 = 120,000 shows this), you are safe.

6. **639 × 196 = 639 × 7 × 7 × 4 = 4473 × 7 × 4 =**
   31,311 × 4 = 125,244

While still holding on to "lines," you now compute the 3-by-3 using the factoring method, to get 125,244. The final step is a simple addition of:

7. 125,244 + "lines" (520,000)

This allows you to say the rest of the answer: ". . . 645,244."

Since a picture is worth a thousand calculations, here's our picture of how this would look:

27,639
× 52,196

"Mom, no knife"

639 × 52 =   33,228
196 × 27 = + 5,292

"Movie lines"

38,520 × 1000 =   38,520,000
52 × 27 × 1 million = + 1,404,000,000

1,442,520,000
639 × 196 = +   125,244

1,442,645,244

I should make a parenthetical note here that I am assuming in doing 5-by-5s that you can write the problem down on a blackboard or piece of paper. If you can't, you will have to create a mnemonic for each of the four numbers. For example, in the last problem, you could store the problem as:

27,639—"Neck jump"
× 52,196—"Lion dopish"

Then you would multiply "lion" × "jump," "dopish" × "neck," "lion" × "neck," and finally "dopish" × "jump." Obviously this would slow you down a bit, but if you want the extra challenge of not being able to see the numbers, you can still solve the problem.

We will conclude this chapter with one more 5-by-5 multiplication:

$$79,838$$
$$\times\ 45,547$$

The steps are the same as those in the last problem. You start with the harder 3-by-2 and store the answer with a mnemonic:

1. **547 × 79 = 547 × (80 − 1) = 43,760 − 547 = 43,213—"Rome anatomy"**

Then you compute the other 3-by-2.

2. **838 × 45 = 838 × 5 × 9 = 4,190 × 9 = 37,710**

Summing the 3-by-2s you commit the new total to memory.

3.   **43,213 "Rome anatomy"**
   × 37,710
   _____

   80,923 "Face Panama"

4. **79 × 45 = 79 × 9 × 5 = 711 × 5 = 3,555**

The 2-by-2 gives you the first digit of the final answer, which you can say out loud with confidence: **"3 billion . . ."**

5. **555 + "Face" (80) = 635**

The millions digits of the answer involve a carryover from 635 to 636, because "Panama" (923) needs only 77,000 to cause it to carry over, and the 3-by-3 (838 × 547) will easily exceed that figure. So you say: **". . . 636 million . . ."**

The 3-by-3 is computed using the addition method:

6.   **838**
                    ×   547 (540 + 7)
                    _____
   540 × 838 =   452,520 (838 × 9 × 6 × 10)
   7 × 800 = +   5,600
                    _____
                      458,120
   7 × 38 = +   266
                    _____
                      458,386

And in the next step you add this total to "Panama" (923,000):

7.　　**923,000**
　+　458,386
　────────────
　　1,381,386

Since you've already used the 1 in the carryover to 636 million, you just say the thousands: "... 381 **thousand** ... 386," and take a bow!
　This problem may be illustrated the following way:

79,838
× 45,547　　　　**"Rome anatomy"**
───────
547 × 79 =　　43,213
838 × 45 = + 37,710
　　　　───────────　　　　　**"Face Panama"**
　　　　　80,923 × 1,000 =　　　80,923,000
　　　79 × 45 × 1 million = + 3,555,000,000
　　　　　　　　　　　　　────────────
　　　　　　　　　　　　　3,635,923,000
　　　　　838 × 547 = +　　　 458,386
　　　　　　　　　　　　　────────────
　　　　　　　　　　　　　3,636,381,386

Exercises: 5-by-5 Multiplication

①	②	③	④
65,154	**34,545**	**69,216**	**95,393**
× 19,423	**× 27,834**	**× 78,653**	**× 81,822**

9

Mathematical Magic

In this chapter we close *Mathemagics* with a collection of mathematical magic that will buttress many of the skills you've learned in this book, as well as show you the pure fun and entertainment to be found in math, which is the reason I went into the field and what keeps me going. My philosophy is that if you can't do math with a certain amount of playfulness, it tends to become too much like work. So let's do a whole chapter of mathematical play!

PSYCHIC MATH

We'll start with an elementary mathemagical effect. It was the first mathemagical trick I ever learned, and my interest in understanding why the trick worked provided me with an incentive to learn algebra.

Say to a volunteer in the audience, "Think of a number, any number, but to make it easy on yourself," you should say, "think of a 1-digit or 2-digit number." After you've reminded your volunteer that there's no way you could know his number, ask him to:

1. Double the number.
2. Add 12.
3. Divide the total by 2.
4. Subtract the original number.

Then say, "Was the answer you got by any chance the number 6?" Try this one on yourself first and you will see that the sequence always produces the number 6 no matter what number is selected.

Why This Trick Works

The number your volunteer chose is represented below by the letter x. Here are the functions you performed in the order you performed them:

1. $2x$
2. $2x + 12$
3. $(2x + 12) \div 2 = x + 6$
4. $x + 6 - x = 6$

For example, let $x = 15$:

1. $2 \times 15 = 30$
2. $30 + 12 = 42$
3. $42 \div 2 = 21$
4. $21 - 15 = 6$

So, no matter what number your volunteer chooses, the final answer will always be 6. If you repeat this trick, have the volunteer add a different number at step 2. The final answer will be half that number ($2 \times 15 = 30; 30 + 18 = 48; 48 \div 2 = 24; 24 - 15 = 9$).

THE MAGIC 1089!

Along the same lines as the magic tricks we just did is the Magic 1089 trick. Have your audience member take out a piece of paper and pencil and:

1. Secretly write down a 3-digit number in which the first digit is larger than the last digit.
2. Reverse that number and subtract it from the first number. If the result is a 2-digit number, tell her to put a 0 in front of it.
3. Take that answer and add it to the reverse of itself.

At the end of this sequence your answer, 1089, will magically appear, no matter what number you originally chose. For example:

$$851 \quad + \quad 685$$
$$- \ 158 \quad - \ 586$$
$$\overline{693} \qquad \overline{099}$$
$$+ \ 396 \quad + \ 990$$
$$\overline{1089} \qquad \overline{1089}$$

Why This Trick Works

No matter what 3-digit number you or anyone else chooses in this game, the final result will always be 1089. Why? Let abc denote the unknown 3-digit number. Algebraically, this is equal to:

$$100a + 10b + c$$

When you reverse the number and subtract it from the original number you get the number cba, algebraically equal to:

$$100c + 10b + a$$

Upon subtracting cba from abc, you get:

$$\begin{array}{l} 100a + 10b + c \\ - (100c + 10b + a) \\ \hline 100\,(a - c) \ + (c - a) = 99\,(a - c) \end{array}$$

Hence, after subtracting in step 2, we must have one of the following multiples of 99: 099, 198, 297, 396, 495, 594, 693, 792, or 891, each one of which will produce 1089 after adding it to the reverse of itself, as we did in step 3.

MISSING DIGIT TRICKS

Using the number 1089 from the last effect, hand a volunteer a calculator and ask her to multiply 1089 by any 3-digit number she likes, but not to tell you the 3-digit number. (Say she secretly multiplies $1089 \times 256 = 278{,}784$). Ask her how many digits are in her answer. She'll reply, "six."

Next you say: "Call out five of your six digits to me in any order you like. I shall try to determine the missing digit."

Suppose she calls out "2 . . . 4 . . . 7 . . . 8 . . . 8." You correctly tell her that she left out the number 7.

The secret is based on the fact that the digits of any multiple of 9 must add up to a multiple of 9. Since 1089 is a multiple of 9, 1089 times any whole number will also be a multiple of 9. Since the digits called out add to 29, and the next multiple of 9 greater than 29 is 36, our volunteer must have left out a 7 (36 − 29 = 7).

There are more subtle ways to force the volunteer to end up with a multiple of 9. Here are some of my favorites:

1. Have the volunteer randomly choose a 6-digit number, scramble its digits, then subtract the smaller 6-digit number from the larger one. Since we're subtracting two numbers with the same mod sum (indeed, the same sum), the resulting difference will have a mod sum of 0, and hence be a multiple of 9. Then continue as before to find the missing digit.

2. Have the volunteer secretly choose a 4-digit number, reverse the digits, then subtract the larger number from the smaller. (This will be a multiple of 9.) Then multiply this by any 3-digit number, and continue as before.

3. Ask the volunteer to multiply 1-digit numbers randomly until the product is seven digits long. This is not "guaranteed" to produce a multiple of 9, but in practice it will do so at least 90% of the time (the chances are high that the 1-digit numbers being multiplied include a 9 or two 3s or two 6s, or a 3 and a 6). I often use this method in front of mathematically advanced audiences who might see through the other methods.

There is one problem to watch out for. Suppose the numbers called out add to a multiple of 9 (say 36). Then you have no way of determining whether the missing digit is 0 or 9. How do you remedy that? Simple, you cheat! You merely say: "You didn't leave out a 0, did you?" If he did leave out a 0, you have completed the trick successfully. If he did not leave out the 0, you say: "Oh, it seemed as though you were thinking of nothing! You didn't leave out a 1, 2, 3, or 4, did you?" He'll shake his head, or say "no." Then you follow with, "Nor did you leave out a 5, 6, 7, or 8, either. You left out the number 9, didn't you?" He'll respond in the affirmative, and you will receive your well-deserved applause!

LEAPFROG ADDITION

This trick combines a quick mental calculation with an astonishing prediction. Handing the spectator a card with ten lines, numbered 1 through 10, have the spectator think of two positive numbers between 1 and 20, and enter them on lines 1 and 2 of the card. Next have the spectator write the sum of lines 1 and 2 on line 3, then the sum of lines 2 and 3 on line 4, and so on as illustrated below.

1	9
2	2
3	11
4	13
5	24
6	37
7	61
8	98
9	159
10	257

Finally, have the spectator show you the card. At a glance, you can tell him the sum of all the numbers on the card. For instance, in our example, you could instantly announce that the numbers sum to 671, faster than the spectator could do using a calculator! As a kicker, hand the spectator a calculator, and ask him to divide the number on line 10 by the number on line 9. In our example, the quotient $\frac{257}{159} = 1.616$. Have the spectator announce the first three digits of the quotient, then turn the card over. He'll be surprised to see that you've already written the number 1.61!

Why This Trick Works

To perform the quick calculation, you simply multiply the number on line 7 by eleven. Here, $61 \times 11 = 671$. The reason this works is illustrated in the table below. If we denote the numbers on lines 1 and 2 by x and y respectively, then the sum of lines 1 through 10 must be $55x + 88y$, which equals 11 times $(5x + 8y)$, that is eleven times the number on line 7. As for the prediction, we exploit the fact that for any positive numbers, a, b, c, d, if $\frac{a}{b} < \frac{c}{d}$, then $\frac{a}{b} < \frac{a+c}{b+d} < \frac{c}{d}$. Thus the quotient of line 10 divided by line 9, $\frac{21x + 34y}{13x + 21y}$, must lie between $\frac{21x}{13x} = \frac{21}{13} \approx 1.615$ and $\frac{34y}{21y} = \frac{34}{21} \approx 1.619$. In fact, if you continue the leapfrog process indefinitely, the ratio of consecutive terms gets closer and closer to $\frac{1 + \sqrt{5}}{2} \approx 1.618$, a number with so many beautiful and mysterious properties that it is often called the "golden ratio."

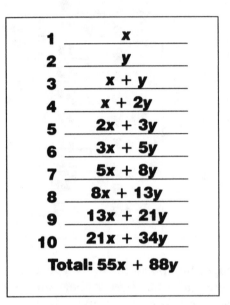

1	x
2	y
3	x + y
4	x + 2y
5	2x + 3y
6	3x + 5y
7	5x + 8y
8	8x + 13y
9	13x + 21y
10	21x + 34y

Total: 55x + 88y

MAGIC SQUARES

Are you ready for a challenge of a different sort? Below you will find what is called a "magic square." There has been much written on magic squares and how to construct them, going back as far as ancient China. Here we describe a way to present magic squares in an entertaining fashion. This is a routine I've been doing for years.

I bring out a business card with the following picture on the back:

8	11	14	1
13	2	7	12
3	16	9	6
10	5	4	15

= 34

I say, "This is called a magic square. In fact, it's the smallest 4-by-4 magic square using the numbers 1 through 16. You'll notice that every row and every column of numbers adds up to the same number—34. Now, I've done such an extensive study of magic squares that I propose to create one for you right before your very eyes."

I then ask someone from the audience to give me any number larger than 34. Let's suppose he says 67.

I then bring out another business card and draw a blank 4-by-4 grid, and place the number 67 to the right of it. Next I ask her to point to the empty squares one at a time, in any order. As she does so, I immediately write a number inside. The end result looks like this:

16	19	23	9
22	10	15	20
11	25	17	14
18	13	12	24

= 67

I continue: "Now with the first magic square, every row and column added to 34. (I usually put this card away at this point.) Let's see how we did with your square."

After checking that each row and column adds up to 67, I say: "But I did not stop there. For you, I decided to go one step farther. Notice that both diagonals also add to 67!" Then I point out that the four squares in the upper left corner sum to 67 (16 + 19 + 22 + 10 = 67), as do the other three four-square corners, the four squares in the middle, and the four corner squares! "They all sum to 67. But don't take my word for any of this. Please keep this magic square as a souvenir from me—and check it out for yourself!"

How to Construct the Magic Square

The key to constructing a magic square using any number is the magic square we bring out initially that sums to 34. Keep that square within eyeshot while you construct the volunteer's magic square. As you draw the 4-by-4 grid, mentally perform the calculations of steps 1 and 2:

1. Subtract 34 from the given number (e.g., 67 − 34 = 33).
2. Divide this number by 4 (e.g., 33 ÷ 4 = 8 with a remainder of 1). The quotient is the first "magic" number. The quotient plus the remainder is the second "magic" number (here our magic numbers are 8 and 9).
3. When the volunteer points to a square, inconspicuously look at the 34-square and see what is in the corresponding square. If it is a 13, 14, 15, or 16, add the second magic number to it (e.g., 9). If not, add the first magic number (e.g., 8).
4. Insert the appropriate number until the magic square is completed.

Please note that when the given number, minus 34, is a multiple of 4 (i.e., it divides by 4 with no remainder), your first and second magic numbers will be the same, so you'll have just one magic number to add to the numbers in your 34-square.

Why This Trick Works

The reason this method works is based on the fact that every group of four squares in the 34-square adds to 34. Suppose the given number had been 82. Since 82 − 34 = 48 (and 48 ÷ 4 = 12), we would add 12 to each square. Then every group of four that use to add to 34 would now add to 34 + 48 = 82. On the other hand, if the given number were 85, our magic numbers would be 12 and 15, so we'd be adding 3 more to the squares showing 13, 14, 15, and 16. Since each row, column, and group of four contains exactly one of these numbers, each group of four would now add to 34 + 48 + 3 = 85.

As an interesting piece of mathemagical trivia, let me point out another astounding property of the "famous" 3-by-3 magic square below. Not only do the rows, columns, and diagonals add to 15, but if you treat the rows of the magic square as 3-digit numbers, you can verify on your calculator that $(492)^2 + (357)^2 + (816)^2 = (294)^2 + (753)^2 + (618)^2$. Also, $(438)^2 + (951)^2 + (276)^2 = (834)^2 + (159)^2 + (672)^2$! Magic "squares," indeed!

4	9	2
3	5	7
8	1	6

= 15

QUICK CUBE ROOTS

Ask someone to select a 2-digit number and keep it secret. Then have him cube the number; that is, multiply it by itself twice (using a calculator). For instance, if the secret number is 68 the calculator will display $68 \times 68 \times 68 = 314,432$. Then ask the volunteer to tell you the answer he got. Once he tells you the cube, you can instantly reveal the original (secret) number (the cube root). How?

To calculator cube roots, you need to learn the cubes from 1 to 10:

$$1^3 = 1$$
$$2^3 = 8$$
$$3^3 = 27$$
$$4^3 = 64$$
$$5^3 = 125$$
$$6^3 = 216$$
$$7^3 = 343$$
$$8^3 = 512$$
$$9^3 = 729$$
$$10^3 = 1000$$

Once you have learned these, calculating cube roots is as easy as π. For instance, with our example problem:

What's the cube root of 314,432?

Seems like a pretty tough one to begin with, but don't panic, it's actually quite simple. As usual, we'll take it one step at a time:

1. Look at the magnitude of the thousands number (the numbers to the left of the comma), 314 in this example.

2. Since 314 lies between $6^3 = 216$ and $7^3 = 343$, according to our list of cubes, the cube root lies in the 60s. Hence the first digit of the cube root is 6.

3. To determine the last digit of the cube root note that only the number 8 has a cube that ends in 2 ($8^3 = 512$) so the last digit must be 8.

Therefore, the cube root of 314,432 is 68. Three simple steps and you're there. Notice that every digit, 0 through 9, appears once among the last digits of the cubes. (In fact, the last digit of the cube root is equal to the last digit of the cube of the last digit of the cube! Go figure that one out!)

Now you try one for practice:

What's the cube root of 19,683?

1. 19 lies between 8 and 27 (2^3 and 3^3).

2. Therefore the cube root is 20-something.

3. The last digit of the problem is 3, which corresponds to $343 = 7^3$, so 7 is the last digit. The answer is 27.

Notice that our derivation of the last digit will only work if the original number is the cube of a whole number. For instance, the cube root of 19,684 is 27.0004572 . . . definitely not 24. That's why we included this in our mathematical magic section and not in an earlier chapter. (Besides, the calculation goes so fast it seems like magic!)

SIMPLIFIED SQUARE ROOTS

Square roots can also be calculated easily if you are given a perfect square. For instance, if someone told you that the square of a 2-digit number was 7569, you could immediately tell her that the original number (the square root) is 87. Here's how.

1. Look at the magnitude of the "hundreds number" (the numbers preceding the last two digits), 75 in this example.

2. Since 75 lies between $8^2 = 64$ and $9^2 = 81$, then we know that the square root lies in the 80's. Hence the first digit of the square root is 8. Now there are two numbers whose square ends in 9, $3^2 = 9$, and $7^2 = 49$. So the last digit must be 3 or 7. Hence the square root is either 83 or 87. Which one?

3. Compare the original number with the square of 85 (which we can easily compute as $80 \times 90 + 25 = 7225$. Since 7569 is larger than 7225, the square root is the larger number, 87.

Let's do one more example.

What's the square root of 4761?

Since 47 lies between $6^2 = 36$ and $7^2 = 49$, the answer must be in the 60's. Since the last digit of the square is 1, the last digit of the square root must be 1 or 9. Since 4761 is greater than $(65)^2 = 4225$, the square root must be 69. As with the previous cube root trick, this method can only be applied when the original number given is a perfect square.

AN "AMAZING" SUM

The following trick was given to me by James "The Amazing" Randi, who has used it effectively in his magic. In this effect the magician is able to predict the total of four randomly chosen 3-digit numbers.

To prepare this trick you will need three sets of nine cards each, and a piece of paper with the number 2247 written down on it and then sealed in an envelope. Next, on each of the three sets of cards do the following:

On Set A write the following numbers, one number on each card:
4286   5771   9083   6518   2396   6860   2909   5546   8174

On Set B write the following numbers:
5792   6881   7547   3299   7187   6557   7097   5288   6548

On Set C write the following numbers:
2708   5435   6812   7343   1286   5237   6470   8234   5129

Select three people in the audience and give each one a set of cards. Have each of your volunteers randomly pick one of the nine cards they hold. Let's say they choose the numbers 4286, 5792, and 5435. Now, in sequence, have each one call out one digit from the 4-digit number, first person "A," then person "B," and finally person "C." Say they call out the numbers 8, 9, and 5. Write down the numbers 8, 9, and 5 (895) and say, "You must admit that this number was picked entirely at random and could not possibly have been predicted in advance."

Next, have the three people call out a different number from their cards. Say they call out 4, 5, and 3. Write 453 below 895. Then repeat this two more times with their remaining two numbers, resulting in four 3-digit numbers, such as:

A	B	C
8	9	5
4	5	3
2	2	4
6	7	5

2	2	4	7

Next have someone add the four numbers, and announce the total. Then slit open the envelope, reveal your predicted total number, and take a bow!

Why This Trick Works

Look at the numbers in each set of cards and see if you can find anything consistent about them. Each set of numbers sums to the same total. Set A numbers total to 20. Set B numbers total to 23. Set C numbers total to 17. With person C's numbers in the right column totaling to 17 you will always put down the 7 and carry the 1. With person B's numbers totaling to 23, plus the 1, you will always put down the 4 and carry the 2. Finally with A's numbers totaling to 20, adding the 2 gives you a total of 2247!

A DAY FOR ANY DATE

One of the classic feats of mental calculation is to tell people the day of the week they were born given their date of birth. The method is based on the fact that every year the dates shift forward by one day except for leap years when they shift by two. Here are the steps involved:

1. Take the last two digits of the year in which your volunteer was born.
2. Divide this number by 4, discard any remainder, and add the result to the original number.
3. Add to this total the number corresponding to the month given in Figure 1, on the next page.
4. Add the day of the month.
5. Finally, divide by 7. The remainder tells you the day of the week your volunteer was born by using Figure 2, on the next page.

For example, let's say your volunteer's date of birth is September 8, 1954 (which happens to be co-author Michael Shermer's birthday). Following the steps in the sequence:

1. The last two digits of the year are 54.
2. Divide 54 ÷ 4 = 13 with a remainder of 2. Discard the 2 and add 13 to the original number of 54 to arrive at 67.
3. To 67 add 6, the number corresponding to September in Figure 1.

4. By this point your running total is 73. To 73 add the day of the month, which is 8 in this case, to arrive at 81.
5. Divide 81 ÷ 7 = 11 with a remainder of 4. The number 4 corresponds to Wednesday in Figure 2.

And, in fact, Michael was born on a Wednesday.

FIGURE 1

Month	Add
January	1*
February	4*
March	4
April	0
May	2
June	5
July	0
August	3
September	6
October	1
November	4
December	6

*For leap years, these numbers are 0 and 3.

FIGURE 2

Remainder	Day
1	Sunday
2	Monday
3	Tuesday
4	Wednesday
5	Thursday
6	Friday
0	Saturday

Here are some mnemonics for remembering the month codes:

MNEMONIC TABLE FOR MONTHS

Month	Key	Mnemonic
January	1	The *first* month.
February	4	A C-O-L-D (4 letters) month.
March	4	The L-I-O-N or L-A-M-B month.
April	0	Think of the Os in April FOOl.
May	2	Think of 2 children, a boy and a girl, running around a Maypole.
June	5	June B-R-I-D-E.
July	0	Think of the Os in the bOOm of fireworks!
August	3	It's a H-O-T month.
September	6	Back to S-C-H-O-O-L.
October	1	Think of a witch riding a broomstick, which looks like the number 1.
November	4	On Thanksgiving we have many things to be "thankful 4."
December	6	Birth of C-H-R-I-S-T or holiday S-E-A-S-O-N.

Remember to subtract 1 from the month code for January and February in leap years (i.e., years where the last 2 digits are a multiple of 4). Note: The years 1800 and 1900 are not leap years, but 2000 is a leap year. For instance, January 16, 1964:

$$\begin{array}{rr} \textbf{January 16, 1964} & \textbf{64} \\ \textbf{64} \div \textbf{4} = & + \textbf{16} \\ \hline & \textbf{80} \\ \textbf{January (leap year):} & + \quad \textbf{0} \\ \hline & \textbf{80} \\ \textbf{16th} & + \textbf{16} \\ \hline & \textbf{96} \end{array}$$

96 ÷ 7 = 13 with a remainder of 5.
5 = Thursday.

For dates in the 1800s, adjust your final answer by adding 2. For dates in the 2000s, subtract 1. For example, Charles Darwin and Abraham Lincoln were both born on February 12, 1809:

$$
\begin{array}{rr}
\mathbf{1809} & \mathbf{09} \\
\mathbf{09 \div 4 =} & \mathbf{+\ \ 2} \\
\hline
& \mathbf{11} \\
\textbf{(February)} & \mathbf{+\ \ 4} \\
\hline
& \mathbf{15} \\
\mathbf{12th} & \mathbf{+\ 12} \\
\hline
& \mathbf{27} \\
\mathbf{1800s} & \mathbf{+\ \ 2} \\
\hline
& \mathbf{29}
\end{array}
$$

29 ÷ 7 = 4 with a remainder of 1.
1 = Sunday

Chapter 1 Answers

Answers: 2-Digit Addition

①
$$\begin{array}{r} 23 \\ + 16 \\ \hline \end{array} \xrightarrow{+10} \begin{array}{r} 33 \\ + 6 \\ \hline \end{array} \xrightarrow{+6} = 39$$

②
$$\begin{array}{r} 64 \\ + 43 \\ \hline \end{array} \xrightarrow{+40} \begin{array}{r} 104 \\ + 3 \\ \hline \end{array} \xrightarrow{+3} = 107$$

③
$$\begin{array}{r} 95 \\ + 32 \\ \hline \end{array} \xrightarrow{+30} \begin{array}{r} 125 \\ + 2 \\ \hline \end{array} \xrightarrow{+2} = 127$$

④
$$\begin{array}{r} 34 \\ + 26 \\ \hline \end{array} \xrightarrow{+20} \begin{array}{r} 54 \\ + 6 \\ \hline \end{array} \xrightarrow{+6} = 60$$

⑤
$$\begin{array}{r} 89 \\ + 78 \\ \hline \end{array} \xrightarrow{+70} \begin{array}{r} 159 \\ + 8 \\ \hline \end{array} \xrightarrow{+8} = 167$$

⑥
$$\begin{array}{r} 73 \\ + 58 \\ \hline \end{array} \xrightarrow{+50} \begin{array}{r} 123 \\ + 8 \\ \hline \end{array} \xrightarrow{+8} = 131$$

⑦
$$\begin{array}{r} 47 \\ + 36 \\ \hline \end{array} \xrightarrow{+30} \begin{array}{r} 77 \\ + 6 \\ \hline \end{array} \xrightarrow{+6} = 83$$

⑧
$$\begin{array}{r} 19 \\ + 17 \\ \hline \end{array} \xrightarrow{+10} \begin{array}{r} 29 \\ + 7 \\ \hline \end{array} \xrightarrow{+7} = 36$$

⑨
$$\begin{array}{r} 55 \\ + 49 \\ \hline \end{array} \xrightarrow{+40} \begin{array}{r} 95 \\ + 9 \\ \hline \end{array} \xrightarrow{+9} = 104$$

⑩
$$\begin{array}{r} 39 \\ + 38 \\ \hline \end{array} \xrightarrow{+30} \begin{array}{r} 69 \\ + 8 \\ \hline \end{array} \xrightarrow{+8} = 77$$

Answers: 3-Digit Addition

①
$$\begin{array}{r} 242 \\ + 137 \\ \hline \end{array} \xrightarrow{+100} \begin{array}{r} 342 \\ + 37 \\ \hline \end{array} \xrightarrow{+30} \begin{array}{r} 372 \\ + 7 \\ \hline \end{array} \xrightarrow{+7} = 379$$

②
$$\begin{array}{r} 312 \\ + 256 \\ \hline \end{array} \xrightarrow{+200} \begin{array}{r} 512 \\ + 56 \\ \hline \end{array} \xrightarrow{+50} \begin{array}{r} 562 \\ + 6 \\ \hline \end{array} \xrightarrow{+6} = 568$$

③
$$\begin{array}{l} 635 \\ +\ 814 \end{array} \xrightarrow{+\ 800} \begin{array}{l} 1435 \\ +\ \ 14 \end{array} \xrightarrow{+\ 10} \begin{array}{l} 1445 \\ +\ \ \ 4 \end{array} \xrightarrow{+\ 4} = 1449$$

④
$$\begin{array}{l} 457 \\ +\ 241 \end{array} \xrightarrow{+\ 200} \begin{array}{l} 657 \\ +\ 41 \end{array} \xrightarrow{+\ 40} \begin{array}{l} 697 \\ +\ \ 1 \end{array} \xrightarrow{+\ 1} = 698$$

⑤
$$\begin{array}{l} 912 \\ +\ 475 \end{array} \xrightarrow{+\ 400} \begin{array}{l} 1312 \\ +\ \ 75 \end{array} \xrightarrow{+\ 70} \begin{array}{l} 1382 \\ +\ \ \ 5 \end{array} \xrightarrow{+\ 5} = 1387$$

⑥
$$\begin{array}{l} 852 \\ +\ 378 \end{array} \xrightarrow{+\ 300} \begin{array}{l} 1152 \\ +\ \ 78 \end{array} \xrightarrow{+\ 70} \begin{array}{l} 1222 \\ +\ \ \ 8 \end{array} \xrightarrow{+\ 8} = 1230$$

⑦
$$\begin{array}{l} 457 \\ +\ 269 \end{array} \xrightarrow{+\ 200} \begin{array}{l} 657 \\ +\ 69 \end{array} \xrightarrow{+\ 60} \begin{array}{l} 717 \\ +\ \ 9 \end{array} \xrightarrow{+\ 9} = 726$$

⑧
$$\begin{array}{l} 878 \\ +\ 797 \end{array} \xrightarrow{+\ 700} \begin{array}{l} 1578 \\ +\ \ 97 \end{array} \xrightarrow{+\ 90} \begin{array}{l} 1668 \\ +\ \ \ 7 \end{array} \xrightarrow{+\ 7} = 1675$$

or

$$\begin{array}{l} 878 \\ +\ 797 \end{array} \xrightarrow{+\ 800} \begin{array}{l} 1678 \\ -\ \ \ 3 \end{array} \xrightarrow{-\ 3} = 1675$$

⑨
$$\begin{array}{l} 276 \\ +\ 689 \end{array} \xrightarrow{+\ 600} \begin{array}{l} 876 \\ +\ 89 \end{array} \xrightarrow{+\ 80} \begin{array}{l} 956 \\ +\ \ 9 \end{array} \xrightarrow{+\ 9} = 965$$

⑩
$$\begin{array}{l} 877 \\ +\ 539 \end{array} \xrightarrow{+\ 500} \begin{array}{l} 1377 \\ +\ \ 39 \end{array} \xrightarrow{+\ 30} \begin{array}{l} 1407 \\ +\ \ \ 9 \end{array} \xrightarrow{+\ 9} = 1416$$

⑪
$$5400 \quad {}^{+\,252} = 5652$$
$$+\ 252 \longrightarrow$$

⑫
$$1800 \quad {}^{+\,855} = 2655$$
$$+\ 855 \longrightarrow$$

⑬
$$6120 \quad {}^{+\,100} \quad 6220 \quad {}^{+\,36} = 6256$$
$$+\ 136 \longrightarrow \quad +\quad 36 \longrightarrow$$

⑭
$$7830 \quad {}^{+\,300} \quad 8130 \quad {}^{+\,48} = 8178$$
$$+\ 348 \longrightarrow \quad +\quad 48 \longrightarrow$$

⑮
$$4240 \quad {}^{+\,300} \quad 4540 \quad {}^{+\,71} = 4611$$
$$+\ 371 \longrightarrow \quad +\quad 71 \longrightarrow$$

Answers: 2-Digit Subtraction

①
$$38 \quad {}^{-\,20} = 18 \quad {}^{-\,3} = 15$$
$$-\ 23 \longrightarrow \quad -\ 3 \longrightarrow$$

②
$$84 \quad {}^{-\,60} = 24 \quad {}^{+\,1} = 25$$
$$-\ 59 \longrightarrow \quad +\ 1 \longrightarrow$$

③
$$92 \quad {}^{-\,40} = 52 \quad {}^{+\,6} = 58$$
$$-\ 34 \longrightarrow \quad +\ 6 \longrightarrow$$

④
$$67 \quad {}^{-\,50} = 17 \quad {}^{+\,2} = 19$$
$$-\ 48 \longrightarrow \quad +\ 2 \longrightarrow$$

⑤
$$79 \quad {}^{-\,30} = 49 \quad {}^{+\,1} = 50 \; or \quad 79 \quad {}^{-\,20} = 59 \quad {}^{-\,9} = 50$$
$$-\ 29 \longrightarrow \quad +\ 1 \longrightarrow \qquad -\ 29 \longrightarrow \quad -\ 9 \longrightarrow$$

⑥
$$63 \quad {}^{-\,50} = 13 \quad {}^{+\,4} = 17$$
$$-\ 46 \longrightarrow \quad +\ 4 \longrightarrow$$

⑦
$$51 \quad {}^{-\,30} = 21 \quad {}^{+\,3} = 24$$
$$-\ 27 \longrightarrow \quad +\ 3 \longrightarrow$$

⑧
$$89 \quad ^{-40} = 49 \quad ^{-8} = 41$$
$$- 48 \longrightarrow \quad - 8 \longrightarrow$$

⑨
$$125 \quad ^{-80} = 45 \quad ^{+1} = 46$$
$$- 79 \longrightarrow \quad + 1 \longrightarrow$$

⑩
$$148 \quad ^{-90} = 58 \quad ^{+4} = 62$$
$$- 86 \longrightarrow \quad + 4 \longrightarrow$$

Answers: 3-Digit Subtraction

①
$$583 \quad ^{-200} = 383 \quad ^{-70} = 313 \quad ^{-1} = 312$$
$$- 271 \longrightarrow \quad - 71 \longrightarrow \quad - 1 \longrightarrow$$

②
$$936 \quad ^{-700} = 236 \quad ^{-20} = 216 \quad ^{-5} = 211$$
$$- 725 \longrightarrow \quad - 25 \longrightarrow \quad - 5 \longrightarrow$$

③
$$587 \quad ^{-300} = 287 \quad ^{+2} = 289$$
$$- 298 \longrightarrow \quad + 2 \longrightarrow$$

④
$$763 \quad ^{-500} = 263 \quad ^{+14} = 277$$
$$- 486 \longrightarrow \quad + 14 \longrightarrow$$

⑤
$$204 \quad ^{-200} = 4 \quad ^{+15} = 19$$
$$- 185 \longrightarrow \quad + 15 \longrightarrow$$

⑥
$$793 \quad ^{-400} = 393 \quad ^{-2} = 391$$
$$- 402 \longrightarrow \quad - 2 \longrightarrow$$

⑦
$$219 \xrightarrow{-200} = 19 \xrightarrow{+24} = 43$$
$$-176 \quad\quad +24$$

⑧
$$978 \xrightarrow{-800} = 178 \xrightarrow{+16} = 194$$
$$-784 \quad\quad +16$$

⑨
$$455 \xrightarrow{-400} = 55 \xrightarrow{+81} = 136$$
$$-319 \quad\quad +81$$

⑩
$$772 \xrightarrow{-600} = 172 \xrightarrow{+4} = 176$$
$$-596 \quad\quad +4$$

⑪
$$873 \xrightarrow{-400} = 473 \xrightarrow{+43} = 516$$
$$-357 \quad\quad +43$$

⑫
$$564 \xrightarrow{-300} = 264 \xrightarrow{+72} = 336$$
$$-228 \quad\quad +72$$

⑬
$$1428 \xrightarrow{-600} = 828 \xrightarrow{+29} = 857$$
$$-571 \quad\quad +29$$

⑭
$$2345 \xrightarrow{-700} = 1645 \xrightarrow{+22} = 1667$$
$$-678 \quad\quad +22$$

⑮
$$1776 \xrightarrow{-1000} = 776 \xrightarrow{+13} = 789$$
$$-987 \quad\quad +13$$

Chapter 2 Answers

Answers: 2-by-1 Multiplication

①
```
    82
×    9
   720
+   18
   738
```

②
```
    43
×    7
   280
+   21
   301
```

③
```
    67
×    5
   300
+   35
   335
```

④
```
    71
×    3
   210
+    3
   213
```

⑤
```
    93
×    8
   720
+   24
   744
```

⑥
```
    49        49
×    9     ×    9
   360 or   450
+   81    −    9
   441      441
```

⑦
```
    28
+    4
    80
+   32
   112
```

⑧
```
    53
×    5
   250
+   15
   265
```

⑨
```
    84
×    5
   400
+   20
   420
```

⑩
```
    58
×    6
   300
+   48
   348
```

⑪
```
    97
×    4
   360
+   28
   388
```

⑫
```
    78
×    2
   140
+   16
   156
```

⑬
```
    96
×    9
   810
+   54
   864
```

⑭
```
    75
×    4
   280
+   20
   300
```

⑮
```
    57
×    7
   350
+   49
   399
```

⑯
```
    37
×    6
   180
+   42
   222
```

⑰
```
    46
×    2
─────
    80
+   12
─────
    92
```

⑱
```
    76
×    8
─────
   560
+   48
─────
   608
```

⑲
```
    29
×    3
─────
    60
+   27
─────
    87
```

⑳
```
    64
×    8
─────
   480
+   32
─────
   512
```

Answers: 3-by-1 Multiplication

①
```
   431
×    6
──────
  2400
+  180
──────
  2580
+    6
──────
  2586
```

②
```
   637
×    5
──────
  3000
+  150
──────
  3150
+   35
──────
  3185*
```

③
```
   862
×    4
──────
  3200
+  240
──────
  3440
+    8
──────
  3448
```

④
```
   957
×    6
──────
  5400
+  300
──────
  5700
+   42
──────
  5742
```

⑤
```
   927
×    7
──────
  6300
+  140
──────
  6440
+   49
──────
  6489
```

⑥
```
   728
×    2
──────
  1400
+   40
──────
  1440
+   16
──────
  1456
```

⑦
```
   328
×    6
──────
  1800
+  120
──────
  1920
+   48
──────
  1968
```

⑧
```
   529
×    9
──────
  4500
+  180
──────
  4680
+   81
──────
  4761
```

⑨
```
   807
×    9
──────
  7200
+   63
──────
  7263
```

⑩
```
   587
×    4
──────
  2000
+  320
──────
  2320
+   28
──────
  2348*
```

⑪
```
   184
×    7
──────
   700
+  560
──────
  1260
+   28
──────
  1288
```

⑫
```
   214
×    8
──────
  1600
+   80
──────
  1680
+   32
──────
  1712
```

⑬
```
    757
×     8
─────────
   5600
+   400
─────────
   6000
+    56
─────────
   6056
```

⑭
```
    259
×     7
─────────
   1400
+   350
─────────
   1750
+    63
─────────
   1813
```

⑮
```
    297        or          297
×     8                  ×     8
─────────                ─────────
   1600    300 × 8 =       2400
+   720     −3 × 8 =   −     24
─────────                ─────────
   2320                    2376
+    56
─────────
   2376
```

⑯
```
    751
×     9
─────────
   6300
+   450
─────────
   6750
+     9
─────────
   6759
```

⑰
```
    457
×     7
─────────
   2800
+   350
─────────
   3150
+    49
─────────
   3199
```

⑱
```
    339
×     8
─────────
   2400
+   240
─────────
   2640
+    72
─────────
   2712
```

⑲
```
    134
×     8
─────────
    800
+   240
─────────
   1040
+    32
─────────
   1072
```

⑳
```
    611
×     3
─────────
   1800
+    33
─────────
   1833
```

㉑
```
    578
×     9
─────────
   4500
+   630
─────────
   5130
+    72
─────────
   5202
```

㉒
```
    247
×     5
─────────
   1000
+   200
─────────
   1200
+    35
─────────
   1235*
```

㉓
```
    188
+     6
─────────
    600
+   480
─────────
   1080
+    48
─────────
   1128
```

㉔
```
    968
×     6
─────────
   5400
+   360
─────────
   5760
+    48
─────────
   5808
```

㉕
```
    499        or          499
×     9                  ×     9
─────────                ─────────
   3600    500 × 9 =       4500
+   810     −1 × 9 =   −      9
─────────                ─────────
   4410                    4491
+    81
─────────
   4491
```

㉖
```
    670
×     4
─────────
   2400
+   280
─────────
   2680
```

㉗
```
    429
×     3
───────
   1200
+    60
───────
   1260
+    27
───────
   1287
```

㉘
```
    862
×     5
───────
   4000
+   300
───────
   4300
+    10
───────
   4310*
```

㉙
```
    285
×     6
───────
   1200
+   480
───────
   1680
+    30
───────
   1710
```

㉚
```
    488
×     9
───────
   3600
+   720
───────
   4320
+    72
───────
   4392
```

㉛
```
    693
×     6
───────
   3600
+   540
───────
   4140
+    18
───────
   4158
```

㉜
```
    722
×     9
───────
   6300
+   180
───────
   6480
+    18
───────
   6498
```

㉝
```
    457
×     9
───────
   3600
+   450
───────
   4050
+    63
───────
   4113
```

㉞
```
    767
×     3
───────
   2100
+   180
───────
   2280
+    21
───────
   2301
```

㉟
```
    312          or          312
×     9                    ×     9
───────                    ───────
   2700      9 × 300 =        2700
+    90      9 ×  12 =     +   108
───────                    ───────
   2790                       2808
+    18
───────
   2808
```

㊱
```
    691
×     3
───────
   1800
+   270
───────
   2070
+     3
───────
   2073
```

*With this kind of problem you can easily say the answer out loud as you go.

Answers: 2-Digit Squares

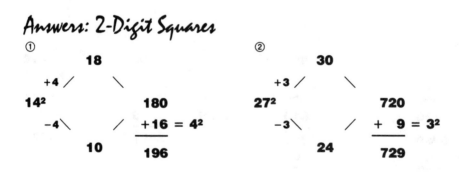

①
```
              18
     +4 /        \
   14²              180
     -4 \        /  +16 = 4²
              10    ────
                    196
```

②
```
              30
     +3 /        \
   27²              720
     -3 \        /  +  9 = 3²
              24    ────
                    729
```

③

$$70$$

$$+5 \nearrow \qquad \searrow$$

$$65^2 \qquad\qquad 4200$$

$$-5 \searrow \qquad \nearrow \quad \underline{+ \ 25} = 25^2$$

$$60 \qquad\qquad 4225$$

④

$$90$$

$$+1 \nearrow \qquad \searrow$$

$$89^2 \qquad\qquad 7920$$

$$-1 \searrow \qquad \nearrow \quad \underline{+ \ \ 1} = 1^2$$

$$88 \qquad\qquad 7921$$

⑤

$$100$$

$$+2 \nearrow \qquad \searrow$$

$$98^2 \qquad\qquad 9600$$

$$-2 \searrow \qquad \nearrow \quad \underline{+ \ \ \ 4} = 2^2$$

$$96 \qquad\qquad 9604$$

⑥

$$32$$

$$+1 \nearrow \qquad \searrow$$

$$31^2 \qquad\qquad 960$$

$$-1 \searrow \qquad \nearrow \quad \underline{+ \ \ \ 1} = 1^2$$

$$30 \qquad\qquad 961$$

⑦

$$42$$

$$+1 \nearrow \qquad \searrow$$

$$41^2 \qquad\qquad 1680$$

$$-1 \searrow \qquad \nearrow \quad \underline{+ \ \ \ 1} = 1^2$$

$$40 \qquad\qquad 1681$$

⑧

$$60$$

$$+1 \nearrow \qquad \searrow$$

$$59^2 \qquad\qquad 3480$$

$$-1 \searrow \qquad \nearrow \quad \underline{+ \ \ \ 1} = 1^2$$

$$58 \qquad\qquad 3481$$

⑨

$$30$$

$$+4 \nearrow \qquad \searrow$$

$$26^2 \qquad\qquad 660$$

$$-4 \searrow \qquad \nearrow \quad \underline{+ \ 16} = 4^2$$

$$22 \qquad\qquad 676$$

⑩

$$56$$

$$+3 \nearrow \qquad \searrow$$

$$53^2 \qquad\qquad 2800$$

$$-3 \searrow \qquad \nearrow \quad \underline{+ \ \ \ 9} = 3^2$$

$$50 \qquad\qquad 2809$$

⑪

$$22$$

$$+1 \nearrow \qquad \searrow$$

$$21^2 \qquad\qquad 440$$

$$-1 \searrow \qquad \nearrow \quad \underline{+ \ \ \ 1} = 1^2$$

$$20 \qquad\qquad 441$$

⑫

$$68$$

$$+4 \nearrow \qquad \searrow$$

$$64^2 \qquad\qquad 4080$$

$$-4 \searrow \qquad \nearrow \quad \underline{+ \ 16} = 4^2$$

$$60 \qquad\qquad 4096$$

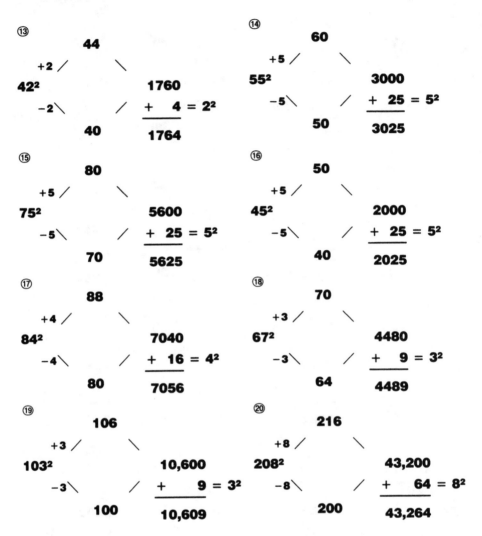

⑬
```
           44
     +2 /      \
  42²              1760
     -2 \      /   +    4 = 2²
           40     1764
```

⑭
```
           60
     +5 /      \
  55²              3000
     -5 \      /   +  25 = 5²
           50     3025
```

⑮
```
           80
     +5 /      \
  75²              5600
     -5 \      /   +  25 = 5²
           70     5625
```

⑯
```
           50
     +5 /      \
  45²              2000
     -5 \      /   +  25 = 5²
           40     2025
```

⑰
```
           88
     +4 /      \
  84²              7040
     -4 \      /   +  16 = 4²
           80     7056
```

⑱
```
           70
     +3 /      \
  67²              4480
     -3 \      /   +   9 = 3²
           64     4489
```

⑲
```
          106
     +3 /      \
  103²              10,600
     -3 \      /   +    9 = 3²
          100     10,609
```

⑳
```
          216
     +8 /      \
  208²              43,200
     -8 \      /   +   64 = 8²
          200     43,264
```

Chapter 3 Answers

Answers: Multiplying by 11

①
$$\begin{array}{r} 35 \\ \times\ 11 \end{array} \rightarrow\ 3\underline{\ \ \ }5 \rightarrow 385$$
$$8$$

②
$$\begin{array}{r} 48 \\ \times\ 11 \end{array} \rightarrow\ 4\underline{\ \ \ }8 \rightarrow 528$$
$$12$$

③
$$\begin{array}{r} 94 \\ \times\ 11 \end{array} \rightarrow\ 9\underline{\ \ \ }4 \rightarrow 1034$$
$$13$$

Answers: 2-by-2 Addition-Method Multiplication Problems

①

$$\begin{array}{r} 31\ (30+1) \\ \times\quad 41 \\ \hline \end{array}$$
$$\begin{array}{rr} 30 \times 41 = & 1230 \\ 1 \times 41 = & +\quad 41 \\ \hline & 1271 \end{array}$$

or

$$\begin{array}{r} 31 \\ \times\quad 41\ (40+1) \\ \hline \end{array}$$
$$\begin{array}{rr} 40 \times 31 = & 1240 \\ 1 \times 31 = & +\quad 31 \\ \hline & 1271 \end{array}$$

②

$$\begin{array}{r} 27\ (20+7) \\ \times\quad 18 \\ \hline \end{array}$$
$$\begin{array}{rr} 20 \times 18 = & 360 \\ 7 \times 18 = & +\ 126 \\ \hline & 486 \end{array}$$

③

$$\begin{array}{r} 59\ (50+9) \\ \times\quad 26 \\ \hline \end{array}$$
$$\begin{array}{rr} 50 \times 26 = & 1300 \\ 9 \times 26 = & +\ 234 \\ \hline & 1534 \end{array}$$

④

$$\begin{array}{r} 53\ (50+3) \\ \times\quad 58 \\ \hline \end{array}$$
$$\begin{array}{rr} 50 \times 58 = & 2900 \\ 3 \times 58 = & +\ 174 \\ \hline & 3074 \end{array}$$

⑤

$$\begin{array}{r} 77 \\ \times\quad 43\ (40+3) \\ \hline \end{array}$$
$$\begin{array}{rr} 40 \times 77 = & 3080 \\ 3 \times 77 = & +\ 231 \\ \hline & 3311 \end{array}$$

⑥

$$\begin{array}{r} 23\ (20+3) \\ \times\quad 84 \\ \hline \end{array}$$
$$\begin{array}{rr} 20 \times 84 = & 1680 \\ 3 \times 84 = & +\ 252 \\ \hline & 1932 \end{array}$$

or

$$\begin{array}{r} 23 \\ \times\quad 84\ (80+4) \\ \hline \end{array}$$
$$\begin{array}{rr} 80 \times 23 = & 1840 \\ 4 \times 23 = & +\quad 92 \\ \hline & 1932 \end{array}$$

⑦
$$62 \ (60 + 2)$$
$$\times \quad 94$$

$60 \times 94 =$ 　5640
$2 \times 94 =$ 　+ 188
　　　　　　　5828

⑧
$$88 \ (80 + 8)$$
$$\times \quad 76$$

$80 \times 76 =$ 　6080
$8 \times 76 =$ 　+ 608
　　　　　　　6688

⑨
$$92 \ (90 + 2)$$
$$\times \quad 35$$

$90 \times 35 =$ 　3150
$2 \times 35 =$ 　+ 　70
　　　　　　　3220

⑩
$$34$$
$$\times \ 11$$
→ 3 __ 4 → 374
　 7

⑪
$$85$$
$$\times \ 11$$
→ 8 __ 5 → 935
　 13

or
　　850
　+ 85
　　935

Answers: 2-by-2 Subtraction-Method Multiplication Problems

①
$$29 \ (30 - 1)$$
$$\times \quad 45$$

$30 \times 45 =$ 　1350
$-1 \times 45 =$ 　− 　45
　　　　　　　1305

②
$$98 \ (100 - 2)$$
$$\times \quad 43$$

$100 \times 43 =$ 　4300
$-2 \times 43 =$ 　− 　86
　　　　　　　　4214

③
$$47$$
$$\times \quad 59 \ (60 - 1)$$

$60 \times 47 =$ 　2820
$-1 \times 47 =$ 　− 　47
　　　　　　　2773

④
$$68 \ (70 - 2)$$
$$\times \quad 38$$

$70 \times 38 =$ 　2660
$-2 \times 38 =$ 　− 　76
　　　　　　　2584

⑤
$$96 \ (100 - 4)$$
$$\times \quad 29$$

$100 \times 29 =$ 　2900
$-4 \times 29 =$ 　− 116
　　　　　　　2784

or
$$96$$
$$\times \quad 29 \ (30 - 1)$$

$30 \times 96 =$ 　2880
$-1 \times 96 =$ 　− 　96
　　　　　　　2784

⑥

		79 (80 − 1)	
	×	**54**	
80 × 54 =		**4320**	
−1 × 54 =	**−**	**54**	
		4266	

⑦

		37
	×	**19 (20 − 1)**
20 × 37 =		**740**
−1 × 37 =	**−**	**37**
		703

⑧

		87 (90 − 3)
	×	**22**
90 × 22 =		**1980**
−3 × 22 =	**−**	**66**
		1914

⑨

		85
	×	**38 (40 − 2)**
40 × 85 =		**3400**
−2 × 85 =	**−**	**170**
		3230

⑩

		57
	×	**39 (40 − 1)**
40 × 57 =		**2280**
−1 × 57 =	**−**	**57**
		2223

⑪

		88
	×	**49 (50 − 1)**
50 × 88 =		**4400**
−1 × 88 =	**−**	**88**
		4312

Answers: 2-by-2 Factoring-Method Multiplication Problems

①
27 × 14 = 27 × 7 × 2 = 189 × 2 = 378 or
14 × 27 = 14 × 9 × 3 = 126 × 3 = 378

②
86 × 28 = 86 × 7 × 4 = 602 × 4 = 2408

③
57 × 14 = 57 × 7 × 2 = 399 × 2 = 798

④
81 × 48 = 81 × 8 × 6 = 648 × 6 = 3888 or
48 × 81 = 48 × 9 × 9 = 432 × 9 = 3888

⑤
56 × 29 = 29 × 7 × 8 = 203 × 8 = 1624

⑥
83 × 18 = 83 × 6 × 3 = 498 × 3 = 1494

⑦
72 × 17 = 17 × 9 × 8 = 153 × 8 = 1224

⑧
85 × 42 = 85 × 6 × 7 = 510 × 7 = 3570

⑨
   $33 \times 16 = 33 \times 8 \times 2 = 264 \times 2 = 528$ or
   $16 \times 33 = 16 \times 11 \times 3 = 176 \times 3 = 528$

⑩
   $62 \times 77 = 62 \times 11 \times 7 = 682 \times 7 = 4774$

⑪
   $45 \times 36 = 45 \times 6 \times 6 = 270 \times 6 = 1620$ or
   $45 \times 36 = 45 \times 9 \times 4 = 405 \times 4 = 1620$ or
   $36 \times 45 = 36 \times 9 \times 5 = 324 \times 5 = 1620$ or
   $36 \times 45 = 36 \times 5 \times 9 = 180 \times 9 = 1620$

⑫
   $48 \times 37 = 37 \times 8 \times 6 = 296 \times 6 = 1776$

Answers: 2-by-2 General Multiplication: Anything Goes!

①

$$
\begin{array}{r}
53 \\
\times \quad 39 \ (40 - 1) \\
\hline
\end{array}
$$

$40 \times 53 = \quad 2120$
$-1 \times 53 = - \quad 53$
$\qquad \qquad \overline{2067}$

or

$$
\begin{array}{r}
53 \ (50 + 3) \\
\times \quad 39 \\
\hline
\end{array}
$$

$50 \times 39 = \quad 1950$
$3 \times 39 = + \ 117$
$\qquad \qquad \overline{2067}$

②

$$
\begin{array}{r}
81 \ (80 + 1) \\
\times \quad 57 \\
\hline
\end{array}
$$

$80 \times 57 = \quad 4560$
$1 \times 57 = + \quad 57$
$\qquad \qquad \overline{4617}$

or $57 \times 81 = 57 \times 9 \times 9 = 513 \times 9 = 4617$

③

$$
\begin{array}{r}
73 \\
\times \ 18 \\
\hline
\end{array}
$$

$73 \times 18 = 73 \times 9 \times 2 = 657 \times 2 = 1314$ or
$73 \times 18 = 73 \times 6 \times 3 = 438 \times 3 = 1314$

④

$$
\begin{array}{r}
89 \ (90 - 1) \\
\times \quad 55 \\
\hline
\end{array}
$$

$90 \times 55 = \quad 4950$
$-1 \times 55 = - \quad 55$
$\qquad \qquad \overline{4895}$ or
$89 \times 55 = 89 \times 11 \times 5 = 979 \times 5 = 4895$

⑤

```
  77    77 × 36 = 77 × 4 × 9 = 308 × 9 = 2772    or
× 36    77 × 36 = 77 × 9 × 4 = 693 × 4 = 2772
```

⑥

```
              92
          ×   53 (50 + 3)
          ─────────
50 × 92 =   4600
 3 × 92 = +  276
          ─────────
            4876
```

⑦

```
   87            90
 × 87        
 ─────    +3 /        \
  87²                    7560
         -3 \      /   +    9 (3²)
              84        ─────────
                         7569
```

⑧

```
             67
         ×   58 (60 − 2)
         ─────────
60 × 67 =   4020
-2 × 67 = −  134
         ─────────
           3886
```

⑨

```
  56    37 × 56 = 37 × 8 × 7 = 296 × 7 = 2072    or
× 37    37 × 56 = 37 × 7 × 8 = 259 × 8 = 2072
```

⑩

```
              59                or              59 (60 − 1)
          ×   21 (20 + 1)                   ×   21
          ─────────                         ─────────
20 × 59 =   1180                60 × 21 =   1260
 1 × 59 = +   59                -1 × 21 = −   21
          ─────────                         ─────────
            1239                              1239
```

```
                           or
      59 × 21 = 59 × 7 × 3 = 413 × 3 = 1239
```

⑪

```
                37
            ×   72 (9 × 8)
            ─────────
37 × 9 × 8 =   333 × 8
           =  2664
```

⑫

```
                      57
                  ×   73 (70 + 3)
                  ─────────
70 × 57 =           3990
 3 × 57 =         +  171
                  ─────────
                    4161
```

⑬

```
                38
            ×   63 (9 × 7)
            ─────────
38 × 9 × 7 =   342 × 7
           =  2394
```

⑭

```
                      43 (40 + 3)
                  ×   76
                  ─────────
40 × 76 =          3040
 3 × 76 =        +  228
                  ─────────
                    3268
```

⑮

$$
\begin{array}{r}
43 \\
\times \quad 75 \ (5 \times 5 \times 3) \\
\hline
\end{array}
$$

$$
\begin{aligned}
43 \times 5 \times 5 \times 3 &= \quad 215 \times 5 \times 3 \\
&= \quad 1075 \times 3 \\
&= \quad 3225
\end{aligned}
$$

⑯

$$
\begin{array}{r}
74 \\
\times \quad 62 \ (60 + 2) \\
\hline
\end{array}
$$

$$
\begin{aligned}
60 \times 74 &= \quad 4440 \\
2 \times 74 &= + \ 148 \\
\hline
&\quad \ \ 4588
\end{aligned}
$$

⑰

$$
\begin{array}{r}
61 \ (60 + 1) \\
\times \quad 37 \\
\hline
\end{array}
$$

$$
\begin{aligned}
60 \times 37 &= \quad 2220 \\
1 \times 37 &= + \quad 37 \\
\hline
&\quad \ \ 2257
\end{aligned}
$$

⑱

$$
\begin{array}{r}
36 \ (6 \times 6) \\
\times \ 41 \\
\hline
\end{array}
$$

$$
\begin{aligned}
41 \times 6 \times 6 &= \quad 246 \times 6 \\
&= 1476
\end{aligned}
$$

⑲

$$
\begin{array}{r}
54 \ (9 \times 6) \\
\times \ 53 \\
\hline
\end{array}
$$

$$
53 \times 9 \times 6 = \quad 477 \times 6 = 2862
$$

⑳

$$
\begin{array}{r}
53 \\
\times \ 53 \\
\hline
56
\end{array}
$$

$$
53^2 \qquad 2800
$$

$$
50 \quad + \qquad 9 \ (3)^2
$$

$$
\overline{\qquad 2809}
$$

㉑

$$
\begin{array}{r}
83 \ (80 + 3) \\
\times \quad 58 \\
\hline
\end{array}
$$

$$
\begin{aligned}
80 \times 58 &= \quad 4640 \\
3 \times 58 &= + \ 174 \\
\hline
&\quad \ \ 4814
\end{aligned}
$$

㉒

$$
\begin{array}{r}
91 \ (90 + 1) \\
\times \quad 46 \\
\hline
\end{array}
$$

$$
\begin{aligned}
90 \times 46 &= \quad 4140 \\
1 \times 46 &= + \quad 46 \\
\hline
&\quad \ \ 4186
\end{aligned}
$$

㉓

$$
\begin{array}{r}
52 \ (50 + 2) \\
\times \quad 47 \\
\hline
\end{array}
$$

$$
\begin{aligned}
50 \times 47 &= \quad 2350 \\
2 \times 47 &= + \quad 94 \\
\hline
&\quad \ \ 2444
\end{aligned}
$$

㉔

$$
\begin{array}{r}
29 \ (30 - 1) \\
\times \ 26 \\
\hline
\end{array}
$$

$$
\begin{aligned}
30 \times 26 &= \quad 780 \\
-1 \times 26 &= - \ 26 \\
\hline
&\quad \ \ 754
\end{aligned}
$$

㉕

$$
\begin{array}{r}
41 \\
\times \ 15 \ (5 \times 3) \\
\hline
\end{array}
$$

$$
41 \times 5 \times 3 = \quad 205 \times 3 = 615
$$

㉖

$$
\begin{array}{r}
65 \\
\times \quad 19 \ (20 - 1) \\
\hline
\end{array}
$$

$$
\begin{aligned}
20 \times 65 &= \quad 1300 \\
-1 \times 65 &= - \quad 65 \\
\hline
&\quad \ \ 1235
\end{aligned}
$$

㉗

$$
\begin{array}{r}
34 \\
\times\ 27\ (9 \times 3) \\
\hline
\end{array}
$$

34 × 9 × 3 =   306 × 3 = 918

㉘

$$
\begin{array}{r}
69\ (70-1) \\
\times\quad 78 \\
\hline
\end{array}
$$

$$
\begin{array}{r}
70 \times 78 = \quad 5460 \\
-1 \times 78 = -\quad 78 \\
\hline
5382
\end{array}
$$

㉙

$$
\begin{array}{r}
95 \\
\times\ 81\ (9 \times 9) \\
\hline
\end{array}
$$

95 × 9 × 9 =   855 × 9 = 7695

㉚

$$
\begin{array}{r}
65\ (60+5) \\
\times\quad 47 \\
\hline
\end{array}
$$

$$
\begin{array}{r}
60 \times 47 = \quad 2820 \\
5 \times 47 = +\ 235 \\
\hline
3055
\end{array}
$$

㉛

$$
\begin{array}{r}
65 \\
\times\ 69\ (70-1) \\
\hline
\end{array}
$$

$$
\begin{array}{r}
70 \times 65 = \quad 4550 \\
-1 \times 65 = -\quad 65 \\
\hline
4485
\end{array}
$$

㉜

$$
\begin{array}{r}
95 \\
\times\ 26\ (20+6) \\
\hline
\end{array}
$$

$$
\begin{array}{r}
20 \times 95 = \quad 1900 \\
6 \times 95 = +\ 570 \\
\hline
2470
\end{array}
$$

㉝

$$
\begin{array}{r}
41\ (40+1) \\
\times\quad 93 \\
\hline
\end{array}
$$

$$
\begin{array}{r}
40 \times 93 = \quad 3720 \\
1 \times 93 = +\quad 93 \\
\hline
3813
\end{array}
$$

Answers: 3-Digit Squares

①

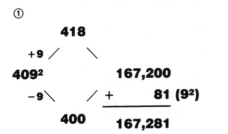

②

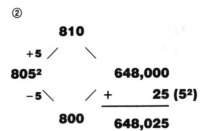

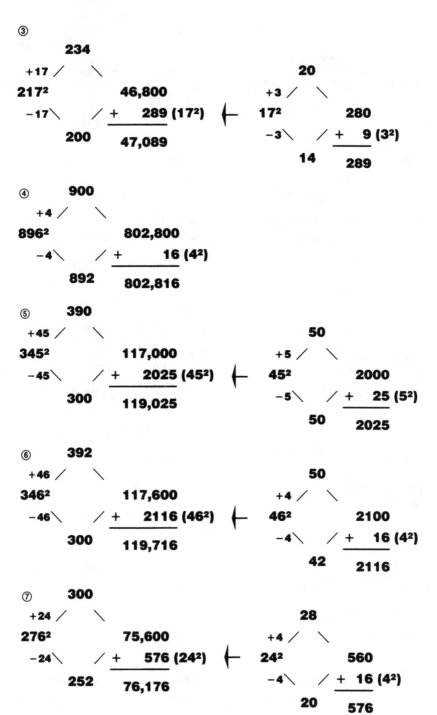

③

```
        234
   +17 /    \
217²           46,800
   -17\    / +    289 (17²)  ⟵
       200    47,089
```

```
                    20
               +3 /    \
          17²           280
             -3\    / +    9 (3²)
                 14    289
```

④

```
        900
    +4 /    \
896²           802,800
    -4\    / +      16 (4²)
       892    802,816
```

⑤

```
        390
   +45 /    \
345²           117,000
   -45\    / +   2025 (45²)  ⟵
       300    119,025
```

```
                    50
               +5 /    \
          45²           2000
             -5\    / +    25 (5²)
                 50    2025
```

⑥

```
        392
   +46 /    \
346²           117,600
   -46\    / +   2116 (46²)  ⟵
       300    119,716
```

```
                    50
               +4 /    \
          46²           2100
             -4\    / +    16 (4²)
                 42    2116
```

⑦

```
        300
   +24 /    \
276²           75,600
   -24\    / +    576 (24²)  ⟵
       252    76,176
```

```
                    28
               +4 /    \
          24²           560
             -4\    / +  16 (4²)
                 20    576
```

⑧

```
        700
 +18 /     \
682²            464,800
  -18\     / +      324 (18²)  ⊢
        664     465,124
```

```
           20
    +2 /     \
   18²           320
    -2\     / +    4 (2²)
        16      324
```

⑨

```
        462
 +32 /     \
431²            184,800
  -31\     / +      961 (31²)  ⊢
        400     185,761
```

```
           32
    +1 /     \
   31²           960
    -1\     / +    1 (1²)
        30      961
```

⑩

```
        800
 +19 /     \
781²            609,600
  -19\     / +      361 (19²)  ⊢
        762     609,961
```

```
           20
    +1 /     \
   19²           360
    -1\     / +    1 (1²)
        18      361
```

⑪

```
        1000
 +25 /     \
975²            950,000
  -25\     / +      625 (25²)  ⊢
        950     950,625
```

```
           30
    +5 /     \
   25²           600
    -5\     / +   25 (5²)
        20      625
```

Chapter 4 Answers

Answers: 1-Digit and 2-Digit Division

①
$$9\overline{)318} \quad 35\tfrac{3}{9}$$
27
48
45
3

②
$$5\overline{)726} \quad 145\tfrac{1}{5}$$
5
22
20
26
25
1

③
$$7\overline{)428} \quad 61\tfrac{1}{7}$$
42
08
7
1

④
$$8\overline{)289} \quad 36\tfrac{1}{8}$$
24
49
48
1

⑤
$$3\overline{)1328} \quad 442\tfrac{2}{3}$$
12
12
12
08
6
2

⑥
$$4\overline{)2782} \quad 695\tfrac{2}{4}$$
24
38
36
22
20
2

⑦
$$17\overline{)738} \quad 43\tfrac{7}{17}$$
68
58
51
7

⑧
$$24\overline{)591} \quad 24\tfrac{15}{24}$$
48
111
96
15

⑨
$$79\overline{)321} \quad 4\tfrac{5}{79}$$
316
5

⑩
$$28\overline{)4268} \quad 152\tfrac{12}{28}$$
28
146
140
68
56
12

⑪
$$11\overline{)7214} \quad 655\tfrac{9}{11}$$
66
61
55
64
55
9

⑫
$$18\overline{)3074} \quad 170\tfrac{14}{18}$$
18
127
126
14

196

Answers: Decimalization

① $\frac{2}{5}$
.40

② $\frac{4}{7}$
.571428

③ $\frac{3}{8}$
.375

④ $\frac{9}{12}$
.75

⑤ $\frac{5}{12}$
.41666

⑥ $\frac{6}{11}$
.5454

⑦ $\frac{14}{24}$
.5833

⑧ $\frac{13}{27}$
.4814

⑨ $\frac{18}{48}$
.375

⑩ $\frac{10}{14}$
.714285

⑪ $\frac{6}{32}$
.1875

Answers: Testing for Divisibility

Divisibility by 2:

① 53428
Yes

② 293
No

③ 7241
No

④ 9846
Yes

Divisibility by 4:

⑤ 3932
Yes

⑥ 67348
Yes

⑦ 358
No

⑧ 57929
No

Divisibility by 8:

⑨ 59366
No

⑩ 73488
Yes

⑪ 248
Yes

⑫ 6111
No

Divisibility by 3:

⑬ 83671
No: 8 + 3 + 6 + 7 + 1 = 25

⑭ 94737
Yes: 9 + 4 + 7 + 3 + 7 = 30

⑮ 7359
Yes: 7 + 3 + 5 + 9 = 24

⑯ 3267486
Yes: 3 + 2 + 6 + 7 + 4 + 8 + 6 = 36

Divisibility by 6:

⑰ 5334
Yes: 5 + 3 + 3 + 4 = 15

⑱ 67386
Yes: 6 + 7 + 3 + 8 + 6 = 30

⑲
248
No: 2 + 4 + 8 = 14

⑳
5991
No: odd

Divisibility by 9:

㉑
1234
No: 1 + 2 + 3 + 4 = 10

㉒
8469
Yes: 8 + 4 + 6 + 9 = 27

㉓
4425575
No: 4 + 4 + 2 + 5 + 5 + 7 + 5 = 32

㉔
314159265
Yes: 3 + 1 + 4 + 1 + 5 + 9 + 2 + 6 + 5 = 36

Divisibility by 5:

㉕
47830
Yes

㉖
43762
No

㉗
56785
Yes

㉘
37210
Yes

Divisibility by 11:

㉙
53867
Yes: 5 − 3 + 8 − 6 + 7 = 11

㉚
4969
No: 4 − 9 + 6 − 9 = −8

㉛
3828
Yes: 3 − 8 + 2 − 8 = −11

㉜
941369
Yes: 9 − 4 + 1 − 3 + 6 − 9 = 0

Divisibility by 7:

㉝
5784

$5784 \xrightarrow{-14} 5770$

$577 \xrightarrow{-7} 570$

57 No

㉞
7336

$7336 \xrightarrow{+14} 7350$

$735 \xrightarrow{-35} 700$

7 Yes

㉟
875

$875 \xrightarrow{-35} 840$

$84 \xrightarrow{-14} 70$

7 Yes

㊱
1183

$1183 \xrightarrow{+7} 1190$

$119 \xrightarrow{+21} 140$

14 Yes

Divisibility by 17:

㊲
694

$694 \xrightarrow{-34} 660$

66 No

㊳
629

$629 \xrightarrow{+51} 680$

68 Yes

㊴
8273

$8273 \xrightarrow{+17} 8300$

83 No

㊵
13855

$13855 \xrightarrow{+85} 13940$

$1394 \xrightarrow{-34} 1360$

$136 \xrightarrow{+34} 170$

17 Yes

Chapters 5 and 6 Answers

Answers: Addition Guesstimation

Exact:

①
```
  1479
+ 1105
------
  2584
```

②
```
  57,293
+ 37,421
-------
  94,714
```

③
```
  312,025
+  79,419
--------
  391,444
```

④
```
   8,971,011
+
   4,016,367
-----------
  12,987,378
```

Guesstimate:

①
```
  1500  or   1480
+ 1100     + 1100
----       ----
  2600       2580
```

②
```
  57,000  or   57,300
+ 37,000     + 37,400
-------      -------
  94,000       94,700
```

③
```
  310,000  or   312,000
+  80,000     +  79,000
--------      --------
  390,000       391,000
```

④
```
  9 million  or   8.9 million  or   8.97 million
+ 4 million     + 4.0 million     + 4.02 million
----------      -----------       ------------
 13 million      12.9 million      12.99 million
                = 12,900,000      = 12,990,000
```

Exact:	Guesstimate:
$ 2.67	$ 2.50
$ 1.95	$ 2.00
$ 7.35	$ 7.50
$ 9.21	$ 9.00
$ 0.49	$ 0.50
$11.21	$11.00
$ 0.12	$ 0.00
$ 6.14	$ 6.00
$ 8.31	$ 8.50
$47.45	$47.00

Answers: Subtraction Guesstimation

Exact:

①
```
  4926
- 1659
------
  3267
```

②
```
  67,221
-  9,874
-------
  57,347
```

③
```
  526,978
-  42,009
--------
  484,969
```

④
```
  8,349,241
- 6,103,839
----------
  2,245,402
```

Guesstimates:

①
```
  4900
- 1700
------
  3200
```

②
```
  67,000   or   67,200
- 10,000     -   9,900
-------      --------
  57,000        57,300
```

③
```
  530,000   or   527,000
-  40,000     -   42,000
--------      ---------
  490,000        485,000
```

④
```
  8.3 million   or   8.35 million
- 6.1 million     -  6.10 million
-------------     --------------
  2.2 million        2.25 million
```

Answers: Division Guesstimation

Exact:

①
$$7\overline{)4379} = 625.57$$

②
$$5\overline{)23{,}958} = 4791.6$$

③
$$13\overline{)549{,}213} = 42{,}247.15$$

④
$$289\overline{)5{,}102{,}357} = 17{,}655.21$$

⑤
$$203{,}637\overline{)8{,}329{,}483} = 40.90$$

Guesstimates:

①
$$7\overline{)4400} = 630$$

②
$$5\overline{)24{,}000} = 4{,}800$$

③
$$13\overline{)550{,}000} = 42{,}000$$

④
$$\approx 300\overline{)5{,}100{,}000} = 3\overline{)51{,}000} = 17{,}000$$

⑤
$$\approx 200{,}000\overline{)8{,}000{,}000} = 200\overline{)8000} = 40$$

Answers: Multiplication Guesstimation

Exact:

①
```
    98
×   27
──────
  2646
```

②
```
    76
×   42
──────
  3192
```

③
```
    88
×   88
──────
  7744
```

④
```
   539
×   17
──────
  9163
```

⑤
```
   312
×   98
──────
 30,576
```

⑥
```
   639
×  107
──────
 68,373
```

⑦
```
   428
×  313
──────
 133,964
```

⑧
```
   51,276
×     489
────────────
 25,073,964
```

⑨
```
   104,972
×   11,201
───────────────
 1,175,791,372
```

⑩
```
      5,462,741
×       203,413
───────────────────
 1,111,192,535,033
```

Guesstimates:

①
```
   100
×   25
──────
  2500
```

②
```
    78
×   40
──────
  3120
```

③
```
    90
×   86
──────
  7740
```

④
```
   540
×   17
──────
  9180
```

⑤
```
   310
×  100
──────
 31,000
```

⑥
```
   646   or      640
×  100       ×   110
──────      ────────
 64,600       70,400
```

⑦
```
   430
×  310
──────
 133,300
```

⑧
```
   51,000
×     490
────────────
 24,990,000
```

⑨
```
   105,000
×   11,000
──────────────────────────────
 1155 million = 1.155 billion
```

⑩
```
       5.5 million
×  200 thousand
──────────────────────────────
 1100 billion = 1.1 trillion
```

Answers: Square Root Guesstimation

Exact (to 2 decimal places):

① $\dfrac{4.12}{\sqrt{17}}$

② $\dfrac{5.91}{\sqrt{35}}$

③ $\dfrac{12.76}{\sqrt{163}}$

④ $\dfrac{65.41}{\sqrt{4279}}$

⑤ $\dfrac{89.66}{\sqrt{8039}}$

Divide and Average:

① $4\overline{)17}^{\;4.2} \qquad \dfrac{4 + 4.2}{2} = 4.1$

② $6\overline{)35}^{\;5.8} \qquad \dfrac{6 + 5.8}{2} = 5.9$

③ $10\overline{)163}^{\;16.3} \qquad \dfrac{10 + 16.3}{2} = 13.15$

④ $60\overline{)4279}^{\;71} \qquad \dfrac{60 + 71}{2} = 65.5$

⑤ $90\overline{)8039}^{\;89} \qquad \dfrac{90 + 89}{2} = 89.5$

Answers: Columns of Numbers

①

672	6
1367	8
107	8
7845	6
358	7
210	3
+ 916	7
11,475	⑨

②

$ 21.56	5
19.38	3
211.02	6
9.16	7
26.17	7
1.43	7
45.32	6
$334.04	⑤

Answers: Subtracting-on-Paper Exercises

①

75,423	3
− 46,298	2
29,125	①

②

876,452	5
− 593,876	2
282,576	③

③
$$3,249,202 \;\rightarrow\; 4$$
$$-\; 2,903,445 \;\rightarrow\; 9$$
___
$$345,757 \;\rightarrow\; ④$$

④
$$45,394,358 \;\rightarrow\; 5$$
$$-\; 36,472,659 \;\rightarrow\; 6$$
___
$$8,921,699 \;\rightarrow\; ⑧$$

Answers: Square Root Exercises

①
$$\begin{array}{r} 3.8\ 7 \\ \sqrt{15.0000} \end{array}$$
$$3^2 = \quad 9$$
___
$$\quad\quad 600$$
$$68 \times 8 = \quad 544$$
___
$$\quad\quad 5600$$
$$767 \times 7 = \quad 5369$$

②
$$\begin{array}{r} 2\ 2.\ 4\ 0 \\ \sqrt{502.0000} \end{array}$$
$$2^2 = 4$$
___
$$\quad\quad 102$$
$$42 \times 2 = \quad 84$$
___
$$\quad\quad 1800$$
$$444 \times 4 = \quad 1776$$
___
$$\quad\quad 2400$$
$$4480 \times 0 = \quad\quad 0$$

③
$$\begin{array}{r} 2\ 0.\ 9\ 5 \\ \sqrt{439.2000} \end{array}$$
$$2^2 = 4$$
___
$$\quad\quad 039$$
$$40 \times 0 = \quad 0$$
___
$$\quad\quad 3920$$
$$409 \times 9 = \quad 3681$$
___
$$\quad\quad 23900$$
$$4185 \times 5 = \quad 20925$$

④
$$\begin{array}{r} 1\ 9 \text{ exactly} \\ \sqrt{361} \end{array}$$
$$1^2 = 1$$
___
$$\quad\quad 261$$
$$29 \times 9 = 261$$
___
$$\quad\quad 0$$

Answers: Pencil-and-Paper Multiplication

①
$$54 \;\rightarrow\; 9$$
$$\times\;\; 37 \;\rightarrow\; 1$$
___
$$1998 \;\rightarrow\; ⑨$$

②
$$273 \;\rightarrow\; 3$$
$$\times\;\; 217 \;\rightarrow\; 1$$
___
$$59,241 \;\rightarrow\; ③$$

③
$$725 \;\rightarrow\; 5$$
$$\times\;\; 609 \;\rightarrow\; 6$$
___
$$441,525 \;\rightarrow\; ③$$

④

3309 ⟶ 6

× 2868 ⟶ 6

─────────────────

9,490,212 ⟶ ⑨

⑤

52,819 ⟶ 7

× 47,820 ⟶ 3

─────────────────

2,525,804,580 ⟶ ③

⑥

3,923,759 ⟶ 2

× 2,674,093 ⟶ 4

─────────────────

10,492,496,475,587 ⟶ ⑧

Chapter 8

Answers: 4-Digit Squares

①

      1468

+234 ╱   ╲   **"Reach off"**   **268**

1234^2   **1,468,000**   +34 ╱   ╲

−234 ╲   ╱ + **54,756** (234^2) ⊢ **234^2**   **53,600**

      1000   **1,522,756**   −34 ╲   ╱ + **1,156** (34^2)

                                       200   **54,756**

②

      9000

+361 ╱   ╲   **"Lesson"**   **400**

8639^2   **74,502,000**   +39 ╱   ╲

−361 ╲   ╱ + **130,321** (361^2) ⊢ **361^2**   **128,800**

      8278   **74,632,321**   −39 ╲   ╱ + **1,521** (39^2)

                                         322   **130,321**

③

      5624

+312 ╱   ╲   **"Tons"**   **324**

5312^2   **28,120,000**   +12 ╱   ╲

−312 ╲   ╱ + **97,344** (312^2) ⊢ **312^2**   **97,200**

      5000   **28,217,344**   −12 ╲   ╱ + **144** (12^2)

                                         300   **97,344**

④

      10,000

+137 ╱   ╲   **"Nachos"**   **174**

9863^2   **97,260,000**   +37 ╱   ╲

−137 ╲   ╱ + **18,769** (137^2) ⊢ **137^2**   **17,400**

      9726   **97,278,769**   −37 ╲   ╱ + **1,369** (37^2)

                                         100   **18,769**

⑤

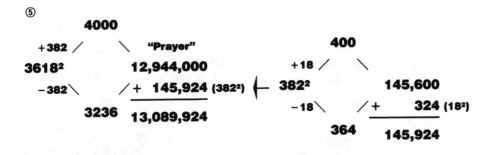

```
          4000
  +382 /        \   "Prayer"                      400
3618²         12,944,000               +18 /        \
  -382\        /+  145,924 (382²) ⟵ 382²           145,600
         3236  13,089,924              -18\        /+      324 (18²)
                                             364       145,924
```

⑥

```
          3000
  +29 /        \ (No mnemonic needed)
2971²         8,826,000
  -29\        /+      841 (29²)
         2942  8,826,841
```

Answers: 3-by-2 Multiplication

①
```
          858
        × 15 (5 × 3)
  858 × 5 ×
3
  = 4290 × 3
  = 12,870
```

②
```
                    796 (800 − 4)
                 ×      19
  800 × 19 =     15,200
   −4 × 19 = −       76
                 15,124
```

③
```
          148                     148 (74 × 2)
        × 62 (60 + 2)  or       ×  62
60 × 148 =   8880        62 × 74 × 2
 2 × 148 = +  296          = 4588 × 2
          9176             = 9176
```

④
```
          773
        × 42 (7 × 6)
773 × 7 × 6
  = 5411 × 6
  = 32,466
```

⑤
```
                    906 (900 + 6)
                 ×      46
900 × 46 =     41,400
  6 × 46 = +      276
                 41,676
```

⑥
$$
\begin{array}{r}
952\ (950 + 2) \\
\times\ \ \ \ 26 \\
\end{array}
$$
950 × 26 = 24,700
2 × 26 = + 52
24,752

⑦
$$
\begin{array}{r}
411\ (410 + 1) \\
\times\ \ \ \ 93 \\
\end{array}
$$
410 × 93 = 38,130
1 × 93 = + 93
38,223

⑧
$$
\begin{array}{r}
967 \\
\times\ \ \ \ 51\ (50 + 1) \\
\end{array}
$$
50 × 967 = 48,350
1 × 967 = + 967
49,317

⑨
$$
\begin{array}{r}
484 \\
\times\ \ \ 75\ (5 \times 5 \times 3) \\
\end{array}
$$
484 × 5 × 5 × 3
= 2,420 × 5 × 3
= 12,100 × 3
= 36,300

⑩
$$
\begin{array}{r}
126\ (9 \times 7 \times 2) \\
\times\ \ \ 87 \\
\end{array}
$$
87 × 9 × 7 × 2
= 783 × 7 × 2
= 5,481 × 2
= 10,962

⑪
$$
\begin{array}{r}
157 \\
\times\ \ \ 33\ (11 \times 3) \\
\end{array}
$$
157 × 11 × 3
= 1727 × 3
= 5181

⑫
$$
\begin{array}{r}
616\ (610 + 6) \\
\times\ \ \ \ 37 \\
\end{array}
$$
610 × 37 = 22,570
6 × 37 = + 222
22,792

⑬
$$
\begin{array}{r}
841 \\
\times\ \ 72\ (9 \times 8) \\
\end{array}
$$
841 × 9 × 8
= 7569 × 8
= 60,552

⑭
$$
\begin{array}{r}
361\ (360 + 1) \\
\times\ \ \ \ 41 \\
\end{array}
$$
360 × 41 = 14,760
1 × 41 = + 41
14,801

⑮
$$
\begin{array}{r}
218 \\
\times\ \ \ \ 68\ (70 - 2) \\
\end{array}
$$
70 × 218 = 15,260
−2 × 218 = − 436
14,824

⑯
$$
\begin{array}{r}
538\ (540 - 2) \\
\times\ \ \ \ 53 \\
\end{array}
$$
540 × 53 = 28,620
−2 × 53 = − 106
28,514

or
$$
\begin{array}{r}
538\ (530 + 8) \\
\times\ \ \ \ 53 \\
\end{array}
$$
530 × 53 = 28,090
8 × 53 = + 424
28,514

⑰
```
                817
          ×      61 (60 + 1)
60 × 817 =   49,020
 1 × 817 = +    817
             49,837
```

⑱
```
             668
          ×   63 (9 × 7)
668 × 9 × 7
= 6,012 × 7
= 42,084
```

⑲
```
              499 (500 − 1)
          ×    25
500 × 25 =   12,500
 −1 × 25 = −     25
             12,475
```

⑳
```
            144
          ×  56 (7 × 8)
144 × 7 × 8
= 1008 × 8
= 8064
```

㉑
```
          281
        ×  44 (11 × 4)
281 × 11 × 4
= 3091 × 4
= 12,364
```
or
```
                  281 (280 + 1)
              ×     44
280 × 44 =    12,320
  1 × 44 = +     44
              12,364
```

㉒
```
            988 (1000 − 12)
          ×  22
1000 × 22 =   22,000
 −12 × 22 = −    264
              21,736
```

㉓
```
          383
        ×  49 (7 × 7)
383 × 7 × 7
= 2,681 × 7
= 18,767
```

㉔
```
                  589 (600 − 11)
              ×    87
 87 × 600 =   52,200
−87 × 11 = −    957
              51,243
```

㉕
```
          286
        ×  64 (8 × 8)
286 × 8 × 8
= 2,288 × 8
= 18,304
```

㉖
```
          853
        ×  32 (8 × 4)
853 × 8 × 4
= 6,824 × 4
= 27,296
```

㉗
$$878 \\ \times \ 24 \ (8 \times 3)$$
$878 \times 8 \times 3$
$= 7{,}024 \times 3$
$\quad = 21{,}072$

㉘
$$423 \ (47 \times 9) \\ \times \ 65$$
$65 \times 47 \times 9$
$= 3{,}055 \times 9$
$\quad = 27{,}495$

㉙
$$154 \ (11 \times 14) \\ \times \ 19$$
$19 \times 11 \times 14$
$= 209 \times 7 \times 2$
$\quad = 1463 \times 2$
$\qquad = 2926$

㉚
$$834 \ (800 + 34) \\ \times \quad 34$$
$800 \times 34 = \quad 27{,}200$
$34 \times 34 = + \ 1{,}156$
$\qquad\qquad\ \ 28{,}356$

㉛
$$545 \\ \times \ 27 \ (9 \times 3)$$
$545 \times 9 \times 3$
$= 4{,}905 \times 3$
$\quad = 14{,}715$

㉜
$$653 \ (650 + 3) \\ \times \quad 69$$
$650 \times 69 = \quad 44{,}850$
$3 \times 69 = + \quad 207$
$\qquad\qquad\ \ 45{,}057$

㉝
$$216 \ (6 \times 6 \times 6) \\ \times \ 78$$
$78 \times 6 \times 6 \times 6$
$= 468 \times 6 \times 6$
$\quad = 2{,}808 \times 6$
$\qquad = 16{,}848$

㉞
$$822 \\ \times \quad 95 \ (100 - 5)$$
$100 \times 822 = \quad 82{,}200$
$-5 \times 822 = - \quad 4110$
$\qquad\qquad\ \ 78{,}090$

Answers: 5-Digit Squares

①
$(45{,}795)^2$:

$$795 \ (800 - 5) \\ \times \quad 45$$
$800 \times 45 = \ 36{,}000$
$-5 \times 45 = - \ 225$
$\qquad\qquad 35{,}775 \times 2000 = \quad 71{,}550{,}000$
$\qquad\qquad (45{,}000)^2 = 2{,}025{,}000{,}000$
$\qquad\qquad\qquad\qquad\qquad\ 2{,}096{,}550{,}000$
$\qquad\qquad\qquad (795)^2 = + \qquad 632{,}025$
$\qquad\qquad\qquad\qquad\qquad\ 2{,}097{,}182{,}025$

"Lillies"

800

795   632,000

+   25 (5^2)

790   632,025

②
(21,231)²:

```
            231
          ×  21 (7 × 3)
```

231 × 7 × 3   "Cousin"
= 1617 × 3
    = 4851 × 2000 =   9,702,000
        (21,000)² = 441,000,000
                    ─────────────
                     450,702,000
        (231)² = +   53,361
                    ─────────────
                     450,755,361

```
        262
        | \
     231²  52,400
       | /  +  961 (31²)
     200    53,361
```

③
(58,324)²:

```
             324 (9 × 6 × 6)
           ×  58
```

   58 × 9 × 6   "Liver"
= 522 × 6 × 6
   = 3132 × 6
      = 18,792 × 2,000 =   37,584,000
          (58,000)² = 3,364,000,000
                       ──────────────
                       3,401,584,000
          (324)² = +   104,976
                       ──────────────
                       3,401,688,976

```
        348
        | \
     324²  104,400
       | /  +    576 (24²)
     300    104,976
```

④
(62,457)²:

```
            457
          ×   62 (60 + 2)
```

60 × 457 = 27,420   "Judge Off"
 2 × 457 = + 914
           ─────────
           28,334 × 2000 =   56,668,000
             (62,000)² = 3,844,000,000
                          ──────────────
                          3,900,668,000
             (457)² = +   208,849
                          ──────────────
                          3,900,876,849

```
        500
        | \
     457²  207,000
       | /  + 1,849 (43²)
     414    208,849
```

⑤

$(89,854)^2$:

$$
\begin{array}{r}
854 \\
\times \quad 89 \ (90 - 1) \\
\hline
\end{array}
$$

$90 \times 854 = 76,860$
$-1 \times 854 = - \ 854$

"Stone"
↑

$$
\begin{array}{r}
76,006 \times 2000 = \quad 152,012,000 \\
(89,000)^2 = 7,921,000,000 \\
\hline
8,073,012,000 \\
(854)^2 = + \qquad 729,316 \\
\hline
8,073,741,316
\end{array}
$$

```
        900
       |   \
 854²   727,200
   |   / + 2,116 (46²)
 808    729,316
```

⑥

$(76,934)^2$:

$$
\begin{array}{r}
934 \ (930 + 4) \\
\times \quad 76 \\
\hline
\end{array}
$$

$930 \times 76 = 70,680$
$4 \times 76 = + \ 304$

"Pie Chef"
↑

$$
\begin{array}{r}
70,984 \times 2000 = \quad 141,968,000 \\
(76,000)^2 = 5,776,000,000 \\
\hline
5,917,968,000 \\
(934)^2 = + \qquad 872,356 \\
\hline
5,918,840,356
\end{array}
$$

```
        968
       |   \
 934²   871,200
   |   / + 1,156 (34²)
 900    872,356
```

Answers: 3-by-3 Multiplication

①

$$
\begin{array}{r}
644 \ (640 + 4) \\
\times \quad 286 \\
\hline
\end{array}
$$

$640 \times 286 = \quad 183,040 \ (8 \times 8 \times 10)$
$4 \times 200 = + \qquad 800$

$$
\begin{array}{r}
183,840 \\
4 \times 86 = + \qquad 344 \\
\hline
184,184
\end{array}
$$

or

$$
\begin{array}{r}
644 \ (7 \times 92) \\
\times 286 \\
\hline
\end{array}
$$

$286 \times 7 \times 92$
$= 2,002 \times 92$
$= 184,184$

②

$$
\begin{array}{r}
596 \ (600 - 4) \\
\times \quad 167 \\
\hline
\end{array}
$$

$600 \times 167 = \quad 100,200$
$-4 \times 167 = - \qquad 668$

$$
\begin{array}{r}
\hline
99,532
\end{array}
$$

③

$$
\begin{array}{r}
853 \\
\times \quad 325 \ (320 + 5) \\
\hline
\end{array}
$$

$320 \times 853 = \quad 272,960$
$5 \times 800 = + \quad 4,000$

$$
\begin{array}{r}
276,960 \\
5 \times 53 = + \qquad 265 \\
\hline
277,225
\end{array}
$$

④
$$343 \ (7 \times 7 \times 7)$$
$$\times 226$$

$226 \times 7 \times 7 \times 7$
$= 1582 \times 7 \times 7$
$= 11,074 \times 7$
$= 77,518$

⑤
$$809 \ (800 + 9)$$
$$\times \quad 527$$

$800 \times 527 =$	$421,600$
$9 \times 527 =$	$+ \quad 4,743$
	$426,343$

⑥
$$942 \ (+42)$$
$$\times \quad 879 \ (-21)$$

$900 \times 921 =$	$828,900$
$-21 \times 42 =$	$- \quad 882$
	$828,018$

⑦
$$692 \ (-8)$$
$$\times \quad 644 \ (-56)$$

$700 \times 636 =$	$445,200$
$(-8) \times (-56) =$	$+ \quad 448$
	$445,648$

⑧
$$446$$
$$\times 176 \ (11 \times 8 \times 2)$$

$446 \times 11 \times 8 \times 2$
$= 4,906 \times 8 \times 2$
$= 39,248 \times 2$
$= 78,596$

⑨
$$658 \ (47 \times 7 \times 2)$$
$$\times 468 \ (52 \times 9)$$

$52 \times 47 \times 9 \times 7 \times 2$
$= 2,444 \times 9 \times 7 \times 2$
$= 21,996 \times 7 \times 2$
$= 153,972 \times 2$
$= 307,944$

⑩
$$273 \ (91 \times 3)$$
$$\times 138 \ (46 \times 3)$$

$91 \times 46 \times 9$
$= 4,186 \times 9$
$= 37,674$

⑪
$$824$$
$$\times \quad 206 \quad = (412)^2$$

$400 \times 424 =$	$169,600$
$12 \times 12 =$	$+ \quad 144$
	$169,744$

⑫
$$642 \ (107 \times 6)$$
$$\times 249 \ (83 \times 3)$$

$107 \times 83 \times 18$
$= 8,881 \times 9 \times 2$
$= 79,929 \times 2$
$= 159,858$

⑬
$$783 \ (87 \times 9)$$
$$\times 589$$

$589 \times 87 \times 9$
$= 51,243 \times 9$
$= 461,187$

⑭
$$871 \ (-29)$$
$$\times \quad 926 \ (+26)$$

$900 \times 897 =$	$807,300$
$-29 \times 26 =$	$- \quad 754$
	$806,546$

⑮

$$341$$
$$\times\ 715$$

$7 \times 341 =$ 2387
$3 \times 15 = +$ 45

$2432 \times 100 = 243{,}200$
$41 \times 15 = +$ 615

$243{,}815$

⑯

$$417$$
$$\times\ \ \ \ 298\ (300 - 2)$$

$300 \times 417 =$ $125{,}100$
$-2 \times 417 = -$ 834

$124{,}266$

⑰

$$557$$
$$\times 756\ (9 \times 84)$$

$557 \times 9 \times 84$
$= 5{,}013 \times 7 \times 6 \times 2$
$= 35{,}091 \times 6 \times 2$
$= 210{,}546 \times 2$
$= 421{,}092$

⑱

$$976\ (1000 - 24)$$
$$\times\ \ \ \ 878$$

$878 \times 1000 =$ $878{,}000$
$-878 \times 24 = -$ $21{,}072$

$856{,}928$

⑲

$$765$$
$$\times\ 350\ (7 \times 5 \times 10)$$

$765 \times 7 \times 5 \times 10$
$= 5{,}355 \times 5 \times 10$
$= 26{,}775 \times 10$
$= 267{,}750$

⑳

$$154 \ (11 \times 14)$$
$$\times 423 \ (47 \times 9)$$

$47 \times 11 \times 14 \times 9$
$= 517 \times 9 \times 7 \times 2$
$= 4653 \times 2 \times 7$
$= 9306 \times 7$
$= 65,142$

㉑

$$545 \ (109 \times 5)$$
$$\times \quad 834$$

$100 \times 834 = \quad 83,400$
$9 \times 834 = + \ 7,506$
_____
$\quad 90,906 \times 5 = 454,530$

㉒

$$216 \ (6 \times 6 \times 6)$$
$$\times 653$$

$653 \times 6 \times 6 \times 6$
$= 3918 \times 6 \times 6$
$= 23,508 \times 6$
$= 141,048$

㉓

$$393 \ (400 - 7)$$
$$\times \quad 822$$

$400 \times 822 = \quad 328,800$
$7 \times 822 = - \quad 5754$
_____
$\quad 323,046$

Answers: 5-by-5 Multiplication

①

$$65,154$$
$$\times 19,423$$
_____ "Neck ripple"
$423 \times 65 = \quad 27,495$
$154 \times 19 = \times \ 2,926$
_____

"Mouse round"

$30,421 \times 1000 = \quad 30,421,000$
$65 \times 19 \times 1 \ \text{million} = +1,235,000,000$
_____
$\quad 1,265,421,000$
$154 \times 423 = + \quad 65,142$
_____
$\quad 1,265,486,142$

②

$$34,545$$
$$\times 27,834$$
_____ "Knife mulch"
$834 \times 34 = \quad 28,356$
$545 \times 27 = +14,715$
_____

"Room scout"

$43,071 \times 1000 = \quad 43,071,000$
$34 \times 27 \times 1 \ \text{million} = + \ 918,000,000$
_____
$\quad 961,071,000$
$834 \times 545 = + \quad 454,530$
_____
$\quad 961,525,530$

③

$$69,216$$
$$\times\ 78,653$$
"Roll silk"

$653 \times 69 =\ \ \ \ 45,057$
$216 \times 78 = +\ 16,848$

"Shoot busily"

$61,905 \times 1000 =\ \ \ \ \ \ \ 61,905,000$
$69 \times 78 \times 1\ \text{million} = +5,382,000,000$

$$5,443,905,000$$
$216 \times 653 = +\ \ \ \ \ \ \ \ 141,048$

$$5,444,046,048$$

④

$$95,393$$
$$\times\,81,822$$
"Cave soups"

$822 \times 95 =\ \ \ \ 78,090$
$393 \times 81 = +31,833$

"Tossup Panama"

$109,923 \times 1000 =\ \ \ \ \ \ \ 109,923,000$
$95 \times 81 \times 1\ \text{million} = +7,695,000,000$

$$7,804,923,000$$
$393 \times 822 = +\ \ \ \ \ \ \ \ 323,046$

$$7,805,246,046$$

Bibliography

Rapid Calculation

Cutler, Ann, and McShane, Rudolph. 1960. *The Trachtenberg Speed System of Basic Mathematics.* New York: Doubleday & Company.

Devi, Shakuntala. 1964. *Figuring: The Joys of Numbers.* New York: Basic Books.

Menninger, K. 1964. *Calculator's Cunning.* New York: Basic Books.

Smith, Steven B. 1983. *The Great Mental Calculators: The Psychology, Methods, and Lives of Calculating Prodigies, Past and Present.* New York. Columbia University Press.

Tirtha, Jagadguru Swami Bharati Krishna, Shankaracharya of Govardhana Pitha. 1965. *Vedic Mathematics or "Sixteen Simple Mathematical Formulae from the Vedas."* Banaras Hindu University Press.

Memory

Lorayne, Harry, and Lucas, Jerry. 1974. *The Memory Book.* New York: Ballantine Books.

Sandstrom, Robert. 1990. *The Ultimate Memory Book.* Los Angeles: Stepping Stone Books.

Recreational Mathematics

Gardner, Martin. 1956. *Magic and Mystery.* New York: Random House.

Gardner, Martin. 1969. *The Unexpected Hanging and Other Mathematical Diversions.* New York: Simon & Schuster.

Gardner, Martin. 1977. *Mathematical Magic Show.* New York: Random House.

Gardner, Martin. 1989. *Mathematical Carnival.* Washington, D.C.: Mathematical Association of America.

Huff, Darrell. 1954. *How to Lie with Statistics.* New York: Norton.

Paulos, John Allen. 1988. *Innumeracy: Mathematical Illiteracy and Its Consequences.* New York: Hill and Wang.

Stewart, Ian. 1989. *Game, Set and Math: Enigmas and Conundrums.* New York: Penguin Books.